'ow.

Making and placing concrete

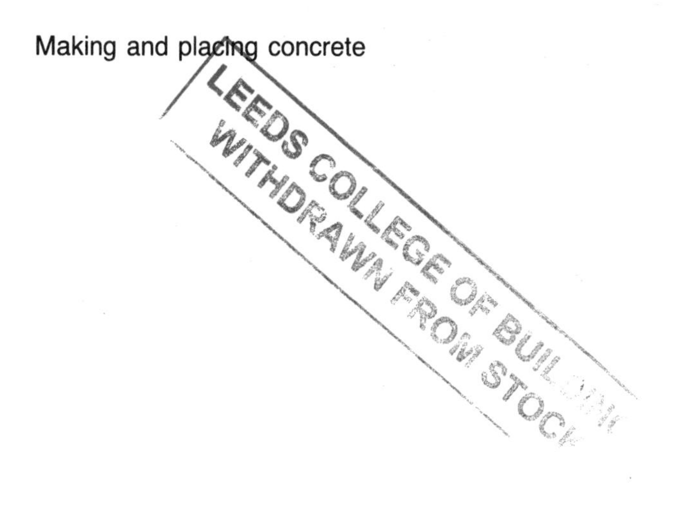

LEEDS COLLEGE OF BUILDING
693.5 BAK
T 12035

LEEDS COLLEGE OF BUILDING LIBRARY
NORTH STREET
LEEDS LS2 7QT
Tel. 0532 430765

Contents

Preface

Concrete is a structural material made from natural ingredients. It is extremely versatile and simple to use but too often, in practice, the basic rules are ignored and an inferior product results. In this book the experience of a working life on site – and in latter years in the lecture theatre – is exploited to describe in practical terms the principles of concrete technology.

The complete process of concrete production is followed from the basic materials employed, to mixing, shaping and placing. Also dealt with are the important aspects of joints, curing, testing and surface finishes. Guidance is also given on the selection and use of admixtures and making good and repairs.

No one book can hope to contain all the information that is needed for every situation. This publication is geared to the man on site who will find in its pages the essential information he requires – the additional reading listed at the end of each chapter will supplement any limitations in coverage that may exist.

I should like to express may grateful thanks to Peter Harlow, Head of Information at The Chartered Institute of Building, for his editorial work and for his support and encouragement, and to the Cement & Concrete Association for its assistance in providing back-up information, photographs and illustrations.

E. M. Baker

Acknowledgements

We are grateful to the following for permission to reproduce illustrations in this book:
The Cement and Concrete Association for Figs. 1.1–1.4, 2.1–2.3, 3.1, 4.1–4.3, 5.1, 5.3, 6.1, 6.2, 6.4–6.6, 7.1, 7.2, 9.4(a)–(d), 9.7, 10.1, 12.1–12.8; Neagron (Construction) Ltd. for Figs. 6.3(a) and (b).

1

Materials

Two of the three main constituents of concrete, i.e. cement and aggregate, are dealt with in this chapter. The significance of water – the third main constituent – will be discussed throughout the book. Admixtures, which may be added to impart a specific property to the concrete, are considered in Chapter 11.

Cement

Cement is the most important constituent (and the most expensive) of concrete and care in its storage, measurement and selection is essential. Portland cement is manufactured from a mixture of shale and limestone, or chalk and clay which are ground together in a slurry and burnt at a very high temperature in a rotary kiln to form a clinker. To this clinker is added a small amount of gypsum – to regulate the setting time – and is then ground to a fine powder. 'Portland' was chosen to describe cement because the finished product resembled Portland stone.

A wide range of Portland cements can be produced either by adding other ingredients or by changing grinding intensities. For example, finer ground cement will harden more quickly. It is clear, therefore, that users must take care that the cement is ordered as specified and that the maker's instructions are followed.

Types of cement

(a) ordinary Portland cement;
(b) rapid-hardening Portland cement;
(c) sulphate-resisting Portland cement;
(d) Portland blast-furnace cement;
(e) white Portland cement;
(f) low-heat Portland cement;

(g) masonry cement;
(h) high alumina cement;
(i) cements manufactured for special purposes.

Ordinary Portland cement (OPC)

OPC is the commonest form of cement, being used for a wide variety of concrete products from paving flags to roads, bridges and oil platforms. It is a reliable, consistent material when manufactured to the current British Standard and it is readily available in all parts of the British Isles. When used in concrete or in mortars it can be attacked by acids and sulphates which may be present in certain soils or ground waters. Sulphates can also occur in clay bricks. When these elements are present, a special cement must be used or other precautionary measures taken.

Rapid-hardening Portland cement (RHPC)

This cement differs from OPC by being ground more finely. It is this fine grinding that develops strength in the cement more rapidly; both cements reach similar strengths on maturity.

The term rapid-hardening should not be confused with 'quick setting'. Setting and stiffening times are similar with both rapid-hardening and OPC, but after this initial period the rapid-hardening cement gains strength more rapidly. This faster rate of strength gain allows formwork to be struck earlier, thus providing savings either in the quantity of formwork required or in time.

An important characteristic of this cement is that of giving out heat on curing. Rapid-hardening cement produces heat earlier than OPC, thus giving it an advantage during cold weather.

Both types of cement are stored and used in the same way. Should they be inadvertently mixed together on site, they may still be used as OPC but this practice is not to be recommended as rapid-hardening cement is more expensive.

Sulphate-resisting Portland cement

Sulphate-resisting Portland cement is made in the same manner and from similar materials as OPC but the chemical constituents are in different proportions. This provides a better performance in resisting attack from sulphates, but like OPC it is not resistant to acids. It is darker than most other Portland cements.

Sulphate-resisting cement is used mainly in concrete below

ground where sulphates are present in the soil or ground water, and sometimes in concrete exposed to sea water. It is important to appreciate that the durability of concrete and its resistance to chemical attack largely depend on it having a minimum cement content and being dense, impermeable – well mixed, well compacted – in other words, it is good concrete. The use of sulphate cement in poorly mixed and badly compacted concrete will not assist its resistance to attack from chemicals. It is normal, therefore, to have a rich rather than lean mix when sulphate-resisting cement is used.

The strength properties of this cement are similar to those of OPC. It should be stored in a similar manner.

Sulphate-resisting cement produces a little less heat than the other Portland cements and this can be an advantage in mass pours, deep basements and foundations. Advice should be sought before using any admixture with this cement, because some will reduce its resistance to sulphate attack.

Portland blast-furnace cement

Portland blast-furnace cement is produced only in Scotland by mixing finely ground OPC clinker with selected granulated blast furnace slag.

It can be used for the same purposes as OPC but it has a slower development of strength, particularly in cold weather. Consequently, it is not suitable when early removal of formwork is required. However, this slow strength development is useful by reducing heat development in thick concrete sections.

White Portland cement

White Portland cement is based on the careful selection of raw materials to reduce the normal iron content that gives Portland cement its grey colour.

Consequently, white cement is expensive so extra care must be taken in handling and storage and in avoiding contamination when batching and mixing; plant and transporting equipment must be kept clean. The finished concrete also needs extra protection.

White cement is used primarily for decorative purposes as, for example, in precast panels, coping and pavings. Care is needed when curing as it is easily soiled during the early stages and is difficult to clean later. Plastic sheeting is excellent for both curing and protection.

Apart from the extra care needed, white cement is treated in the same way as Portland cement, its setting time and strength being similar.

Surface crazing appears to be more prevalent when the concrete matures in an industrial environment. This is due to the contrast of the darkening and filling of the crazing against the white surface – white cement concrete does not craze any more than grey OPC concrete.

Low-heat Portland cement

Low-heat Portland cement has a low rate of strength development and, as it name implies, it generates less early heat than OPC. These properties are particular useful for dam and other mass concrete construction. This cement is generally made specially for contracts needing more than 300 tonnes of concrete.

Masonry cement

Masonry cement is an OPC with additions of fine inert powder and air-entraining agent as a substitute for lime. This gives this cement a consistent workability for use in mortars for brickwork and blockwork, so reducing the variability of site-mixed mortars based on OPC. It must not be used for concrete and advice should be sought on the mixes needed to suit varying exposure and weather conditions.

Special cements

Special cements include water-repellent and hydrophobic cements. The former is an OPC with a water-repellent additive used mainly on backing coats to renderings. Hydrophobic cement is intended for use in poor storage conditions. Both cements are made to meet special orders.

High alumina cement (HAC)

Limestone and bauxite are fused in a furnace to produce a cement, somewhat darker than ordinary Portland, that stiffens at about the same rate as OPC. Once stiffened, strength develops extremely rapidly (after 24 hours it can reach 55 N/mm^2).

Contamination with Portland cement causes a 'flash set' (accelerated setting) so all mixers, shovels and barrows, etc. must be carefully cleaned to remove any traces of ordinary cement. High alumina cement should always be stored separately in a clearly marked position and admixtures should not be used. HAC's rapid gain in strength is useful, for example in floor repairs of shops, by providing a walking or working surface in a few hours. It is also used in high temperature applications, but its structural use is prohibited. Advice should be sought from the manufacturer before use.

Table 1.1 Typical uses of some cements

Cement type	*Portland*						*Other*	
Use	Ordinary	Rapid-hardening	Blast furnace	Low heat	Sulphate resisting	White	Hydrophobic	High alumina
Ordinary construction	★	★	★			★	★	
Precast work	★	★				★		
Coloured concrete						★		
Exposed environment	★	★			★			
High early strength		★						
Cold weather concreting		★						
Marine environment	★	★			★			
Massive construction				★				
Resistance to attack by sulphates					★			
Refractory (high temperature) concrete								★
Prolonged or tropical cement storage							★	

Cement in use

Hot cement

The heat generated by grinding during manufacture can be retained for a considerable time in bulk cement. Although the cement can still be hot when it reaches the customer, tests have shown that the effect on workability and strength is not significant and a slight improvement of setting time may result, similar to the effect of adding hot water in winter as a frost precaution.

Storing cement

Cement stored in airtight tins will keep indefinitely. Kept in a good silo it will be satisfactory for about 3 months but when in normal three-ply paper bags, even under good conditions, it can lose considerable strength after 4 to 6 weeks. Every effort must be made to exclude not only water but also damp air. Any torn or damp bags should be rejected.

Prolonged storage in damp conditions sometimes results in partial or 'air setting'. A practical test of usability is to crumble the cement between the fingers. If it fails to separate then it should not be used for structural purposes. Cement that has 'air set' but is still usable has probably lost some of its strength. This must be taken into account by either using the cement for 'non-structural' purposes or by taking cube tests to determine what strength can be achieved. Increased concrete strength can be effected by increasing the cement content.

Since cement is the least durable and the most expensive of the materials used to make concrete it should be treated accordingly. Careful storage will more than repay the effort expended.

Methods of storage

There are three main methods of storing cement.

1. in the open;
2. shed;
3. bulk in a silo;

In the open On small jobs, or even large jobs where there is a need to have small amounts of cement available at various points over a large area, for example piling foundations or scattered housing sites, the cement may have to be stored in the open.

A dry base is essential, raised at least 100 mm higher than the surrounding growth. If the surrounding growth exceeds 100 mm, boards or bricks are suitable to raise the base. An improvement is to cover the base with a plastic sheet to prevent rising moisture.

These stacks must be well covered with tarpaulins or plastic sheeting which overlap and which are fixed or weighted down with heavy objects on top and at ground level.

If a cement stack forms a pitch this will assist the discharge of rain. Covering cement out in the open needs to be thorough; if done badly the water will be concentrated at the gaps and the covers will do more harm than good.

Shed Any shed for storing cement must be watertight and have a sound, dry floor. A doubtful floor can be improved with a plastic sheet.

The bags should be stacked away from the walls and not more than eight to ten bags high. They should be so stacked that the first bags in are the first to be used. It is important that the cement is used in the same order as it was delivered.

The bottom bags sometimes appear to have become 'air set' but this may only be compaction from the weight of those above. This cement will easily separate during mixing.

Fig. 1.1 Storage of cement

As draughts can bring in damp air, stacking the bags close together and covering with plastic sheeting are additional precautions.

The hut door should close properly and be kept shut.

Bulk storage – silos There are considerable benefits for the contractor and men on site in storing cement in silos, if a few simple rules are followed (see later) and careful consideration given to size, siting, ordering and maintenance. Advantages are that:

1. Bulk cement is cheaper and the contractor does not have to unload or interrupt planned work for unloading.
2. Bags do not have to be handled to the mixer. Wastage due to broken bags is eliminated.
3. Deterioration is reduced since silos are weatherproof.
4. Silos allow the mixer to produce a full batch rather than having to match the exact number of 50 kg bags of cement.
5. Cleaner working conditions result.

To get best results the storage capacity needs to be related to the:

(a) maximum rate of cement use;
(b) quantity that can be delivered and reliability of deliveries;
(c) site access – road and traffic conditions.

Few silos indicate when they are full – high-level indicators are available and should be used.

As air pressure is used to transfer cement from the road transporter to the silo an allowance of 20 to 25 per cent must be made for the increase in bulk of freshly aerated cement which occupies more space in the silo.

Silos are also fitted with a mechanism to correctly weigh a batch and to discharge this into the mixer. This requires regular maintenance and adherance to the manufacturer's recommendations.

Check list for silo storage of cement

Silo size

- Is there enough cement in hand if a fresh load fails to arrive as scheduled?
- How much cement is needed to maintain production between deliveries?

Fig. 1.2 Most sites will have a portable cement silo – its size related to the concrete production rate to ensure that fresh cement is always used

- What effect will aeration have on silo capacity?
- Is it easy to tell when the silo is full?

Site arrangement

- Is the access road suitable for delivery, bearing in mind the weight and size of the lorry?
- Is there enough space for the delivery tanker to back up to the silo?
- Is there enough headroom for tipping?
- Is the ground sufficiently firm?
- Is the filling pipe correctly positioned?
- Are the pipe runs reasonably short and straight?
- Is the cement type clearly marked on each silo?
- Is the point of cement discharge handy for the mixer driver?

Ordering and delivery of cement

- How much notice of delivery does the supplier need?
- Is it possible to accept a full load, preferably into one silo?
- Does the delivery man need help with reversing?

Maintenance

- Are the air filters regularly cleaned?
- Are the air filters easy and safe to reach – particularly if situated at the top of the silo?
- Does the pressure relief valve operate freely?
- Is the weighing gear correctly calibrated?
- Is the weigh hopper clean?

Aeration

- Is the air supply functioning and are the porous plugs clear?

The cement delivery vehicle is a pressure tanker fitted with an air compressor to assist in discharging its load. The driver controls the mixture of air and cement to carry the cement into the silo. The cement separates on entering the silo, leaving dust-laden air needing filtering before it is allowed to enter the outside atmosphere.

The filters require regular cleaning if they are to work properly; this is normally done by shaking them and using an air blast when necessary: The manufacturer of the silo will specify the correct intervals for carrying out this important task.

Should the filters be allowed to clog, dust will be blown about, creating unpleasant conditions for the site staff and neighbours.

It may also be an offence under the Health and Safety at Work etc Act.

Most silos have a pressure relief valve fitted in the form of cap or plate that lifts off if the pressure of air or cement becomes too great. This simple device can fail if neglected. Hardened cement can build up around it and when it fails a lot of dust may be blown about.

Some silos have these air outlets piped down to filters at ground level to assist regular maintenance, but many have them fitted at the top, access being by a vertical ladder. This ladder should have a cage around it to protect personnel from falling.

Cement replacements

The need to conserve the fuel employed in the manufacture of cement has increased the interest and use of cement-extending materials. Two of these commonly used in the UK are:

(a) blast furnace slag (produced by granulation of the slag obtained from the production of iron);

(b) pulverised fuel ash (PVA) (the product from coal-fired electricity generating stations).

Both of these require the presence of approximately 70–75 per cent of cement. They are usually blended by the cement manufacturers.

The site's interest in these products is the effect on strengths and setting times. Early strengths are reduced but later strengths can increase. Heat of hydration and temperature rise are reduced.

Due to lower heat of hydration, PVA may be useful for large pours. Slower strength development could delay formwork striking times. Blended cements should always be clearly marked and used in accordance with engineer's instructions.

Aggregates

Cement and water alone would make a very expensive structural material, apart from the attendant problems of accommodating the shrinkage that occurs. Aggregates are added, up to 80 per cent by volume in ordinary concrete, to provide bulk. They are also much cheaper than cement.

Aggregates originate from many sources and occur in a wide variety. The desired properties and requirements are fully described in British Standard 882 (1983).

They must be strong, clean and chemically inert, non-laminar in form and suitably graded.

Aggregates are described as 'lightweight', 'normal-weight', or 'heavy'. They may be either 'natural', e.g. crushed or uncrushed gravel, crushed stone and sands derived from crushing or sieving, or 'artificial', e.g. blast-furnace slag or burnt clay, and are described as either 'lightweight' or 'normal-weight'.

The sizes of the pieces comprising an aggregate should vary from the largest specified size to the finest, with middle sizes predominating.

All aggregates contain some shells and/or salt. These are not normally harmful, providing limits are set on the content of either.

It is advisable to seek advice before using an aggregate from a new source of supply for the first time.

Aggregates are graded into sizes of particle and the gravels are washed. Variations occur in the operation of these processes, so the user should have the means and the experience to check by observation and by site tests the cleanliness, grading and moisture content.

Characteristics of aggregates

All aggregates need to be durable and clean. They should be hard and not contain anything that will decompose, or change in volume when exposed to the weather, or affect the reinforcement. Clay, for example, may swell or soften and form weak pockets in the concrete.

It is an essential quality of good aggregates that they are free of organic impurities. Neither should they contain fine dust as this may prevent the cement paste from coating the aggregate properly, preventing bonding and thus lowering the strength of the concrete.

Types of aggregates

Fine aggregates

In the category of fine aggregates are natural sand, crushed rock and crushed gravel that pass a 5 mm BS sieve when dry.

These materials vary considerably in character, depending on their location and methods of extraction and grading and may be described as 'sharp' or 'soft'. 'Sharp' sand has angular grains and is

Fig. 1.3 Three types of sand; shapes and sizes of aggregates

used mainly for concrete, whereas 'soft' sand has smaller rounded grains and is used for mortars and renderings. They are known as 'concreting' sand and 'building' sand respectively. Concreting sand is classified as coarse, medium or fine (C, M or F).

Sand provides the predominant colour effect in concrete. Since it can vary in different areas of the same pit or working, special care must be taken if a consistent colour is required.

Coarse aggregates

Coarse aggregate describes the material that is retained on a 5 mm sieve when dry. The maximum size of coarse aggregate used depends on the application of the concrete. Reinforcement in concrete has to be surrounded thoroughly, so 20 mm is the maximum size normally employed. There are advantages where larger-sized aggregate can be used as the overall surface area is smaller, thereby requiring less cement paste. For example, a roughly cube-shaped stone, if split in two, will have its surface area increased by one-third. This will require additional cement paste to cover.

The mix designer has to select the appropriate aggregate to suit the application; 10 mm may be required for small section work with a lot of close reinforcement, 20 mm for general work and 100–150 mm for large unreinforced pours.

Graded aggregates

Natural aggregates, whether obtained from gravel pits on the sea bed, will consist of different size stones randomly mixed. This variation will affect both workability and strength of the concrete from batch to batch. Where these are important the coarse material can be obtained within a specific size or accurately proportioned at the source of supply. This is usually the concern of the mix designer or structural engineer rather than the site user.

'All-in' aggregate

All-in aggregate is a mixture of coarse and fine aggregate either as extracted or delivered. It is not allowed for structural concrete work because the grading is likely to vary considerably. Consequently, its use is confined to low grade concretes such as those used for pipe bases, protection or mass concrete for non-structural applications.

Water content

Most aggregates will contain a certain amount of water which varies according to the weather. Sand will retain more water than coarse aggregates but this is not easy to control. It is helpful if the base of the stockpiles has a fall to allow drainage. Aggregates should be left to drain for at least half a day before use and then taken from the top of the pile as work proceeds.

On small jobs the mixer driver needs to watch constantly for the effect of surplus water; at times sand can contain up to 15 per cent moisture. On larger projects control is the responsibility of the quality control engineer, details of which are given in subsequent chapters.

Site control of aggregates

The type of aggregate to be used on a particular job will be selected, specified and approved before concreting begins.

At the earliest opportunity a representative from the site should visit the supplier to acquaint himself with the personnel and to assess the resources, limitations and standard of efficiency that can be expected throughout the contract.

Because of the variations that occur in aggregates each load must be carefully examined, preferably before tipping, to check if the grading is consistent and as specified (grading processes do fail at the suppliers). Any excess of large or small aggregates can be seen and prevented from being tipped on to the stockpile.

A sample of aggregate can be rubbed between hands to see if soil or silt remains on them. If this occurs then further tests should

Fig. 1.4 Typical batching and mixer set-up with drag-line skip for loading hopper. The weighing dial is in full view and can be easily read by both the mixer driver and the dragline skip operator

be made to check if this material is acceptable. Appropriate tests are described in Chapter 12. Any doubts about gradings, shape or cleanness should be reported to the site engineer. It is also advisable to check delivery tickets against capacity of vehicle and occasionally to have a delivery weighed as a precaution against fraud or error. Care must be taken to provide, and to maintain, robust partitions to separate the different aggregates in use. A base should be laid to drain towards the outside of the stockpiles and be continued away from the mixer to prevent wheeled lorries taking mud to the stored materials.

Stockpiles should be well filled and aggregates taken to the mixer from the top to ensure that they have drained and to avoid scraping up mud or segregated materials. They should also be kept clean and free from canteen and other site rubbish and away from trees.

It is useful to have covers of tarpaulins or plastic sheeting to use for protection if immediate concreting is not anticipated or in winter. Further information about winter working can be found in Chapter 11.

Lightweight aggregates

As an alternative to the heavy natural aggregates there are lightweight aggregates, such as pumice (volcanic lava). This material is imported into Great Britain from Italy and Sicily with an 'all-in' grading of 20 mm down.

Another natural lightweight aggregate is diatomite, which is a calcined diatomaceous rock formed from minute fossil skeletons of marine life. This is commercially available in the United Kingdom and is found principally in the Isle of Skye and in Cumbria.

The site user must know the properties of these natural lightweight aggregates for, in use, they behave very differently from ordinary aggregates. They can, however, be mixed in standard batching plants and concrete mixers.

Such natural lightweight aggregates are expensive, so most of those used in the UK are manufactured.

Types of lightweight aggregate

Aglite Aglite is produced by mixing a blend of clay and shale with ground coke and passing it over a sinter-strand hearth and then through crushers to give it an angular shape. It is available in three sizes: 15–10 mm, 10–5 mm and 5 mm down.

Leca Leca is produced from a special grade of clay, suitable for bloating, which is ground and pre-treated before passing through a rotary kiln; this process creates a range of smooth spherical pellets with a glazed but porous skin. It is available in three sizes: 20–10 mm, 10–3 mm and 3 mm down.

Sintered pulverised-fuel ash (pfa)–Lytag Ash collected from the flue gases discharged from modern power stations burning pulverised fuel is known as 'pulverised-fuel ash', 'fly-ash' or 'pfa' and consists of particles which are as fine as, if not finer than, cement. Lightweight aggregate is produced by dampening the pfa with water and mixing it with coal slurry in screw mixers. The material is then fed on to rotating pans (known as pelletisers) to form spherical pellets which are then sintered at a temperature of 1200 °C. This causes the ash particles to coalesce, without fully melting, to form a lightweight aggregate. Known as Lytag, the aggregate is available in three sizes: 12–8 mm and 8–5 mm, which are spherical, and fine–5 mm down–which is angular since it is made by crushing the larger grades.

Exfoliated vermiculite Vermiculite is a mineral of laminar structure which exfoliates rapidly when heated, thereby radically reducing its density.

The raw ore, usually imported from America, Australia or South Africa, is first dried, crushed and then graded for size. The grading process is carried out by the age-old method of 'winnowing' in a stream of air. The individual grades of material are then rapidly passed through hot burners (about 1000 °C); this causes the vermiculite to exfoliate, the formation of steam forcing the laminae apart.

Exfoliated vermiculite is available in four standard grades, but for concrete the recommended sizes are 7–6 mm and 6–5 mm.

Expanded perlite Perlite, a glassy volcanic rock containing water, is imported from Italy, Greece and America. It is also found in Northern Ireland. The industrial process of expanding the rock to form a lightweight aggregate consists of crushing the material to graded sizes and rapidly heating it to the point of incipient fusion (approximately 1800 °C). At this temperature, the water dissociates and expands the glass into a balloon-like formation of small bubbles to produce a cellular material with a correspondingly low density.

Expanded perlite is available in four grades, but for concrete the recommended grading is within Zone L1 of BS 3797; i.e. maximum size 6 mm.

Wood particles Graded wood particles have been used as an aggregate for many years, but some method of pre-treatment of the material is usually necessary – otherwise the tannins, soluble carbohydrates, waxes and resins which wood contains may affect the hydration of the cement.

Various patent processes are available, but with most softwoods it is merely necessary to mix in calcium chloride or lime. Another treatment is to boil the particles in water to which ferrous sulphate has been added.

Plastic particles Expandible plastic particles, such as polystyrene, take the form of plastic beads in which an expanding agent has been dissolved. The plastic is softened in steam heat and the agent forces the beads to expand to about four times their original diameter. The beads after expansion may be coated with resin and cement.

The maximum size of expanded polystyrene beads is about 4 mm.

Natural lightweight aggregates

Most lightweight aggregates, because of their porosity, need pre-wetting when mixing and it is advisable to add the cement last to reduce the possibility of losing cement paste in the aggregate. Mixing times are critical, vibration should not be prolonged and curing is more important than for concrete with normal aggregates.

Advantages of using lightweight aggregates

Apart from the overall benefit of using waste materials, e.g. pfa, the lightweight aggregates reduce the need to exploit natural resources and the tying up of agricultural land.

The other main advantages to be gained by using lightweight aggregates are:

1. A low bulk density providing a high thermal insulation; this is particularly valuable now buildings have to comply to tighter building regulations aimed at heat conservation.

2. A low density which is particularly useful to designers of elevated structures, i.e. bridges or large, tall buildings as its use reduces the self-load and permits economies in structural sections and foundations.
3. A better fire resistance than many other materials.
4. Easier transportation and handling when used for large precast units.
5. Lightness and ease of handling of both loadbearing and non-loadbearing concrete block.
6. Good insulation to roof screeds without adding very much to the overall load.

Summary

Effective supervision at site level is dependent on those using the constituent materials of concrete being able to identify them. By examination or testing it is possible to establish if the materials are of the quantity, quality and cleanness required. Most aggregates and cement – even the water – have variations which are acceptable. On the other hand, badly graded aggregates, some grades and types of sand and dirty, polluted water, will affect the performance of the concrete.

As materials vary according to their source, local advice should be sought before operating in a new area or using a new supplier.

Suggested reading

1. Cement & Concrete Association. *Concrete practice.* 1979.
2. Cement & Concrete Association. *Cements* (Man on the job leaflet).
3. Watson, R. V. *Construction Guide: storing cement in silos on site* (2nd edn). Cement & Concrete Association, 1979.
4. Spratt, B. H. *An Introduction to Lightweight Concrete* (6th edn). Cement & Concrete Association, 1980.

2

Mixing

Mix design

Cement starts to hydrate immediately it is mixed with water. The hardening and strength development are the result of a chemical reaction and not a drying out process: each particle of cement develops a growth on its surface which gradually spreads and links with similar growth from other particles resulting in stiffening, hardening and gain of strength.

This stiffening or setting, as it is sometimes called, usually occurs in the first hour or two but is dependent on a number of variable factors, i.e. mix proportions, weather conditions, etc. The chemical reaction is almost complete after about one month but concrete continues to gain strength slowly for many years. This 'hydration' generates heat which can be of benefit to the user in cold weather conditions.

Once the correct, specified materials are on site the next most important step in the production of concrete is to mix these together thoroughly in the correct proportions.

The objectives of mix design are to produce a satisfactory end product, e.g. beam, column or slab, and to utilise the nearest available suitable materials and manpower as economically as possible.

The mix designer will have access to information in the form of tables, charts, grading curves and water/cement ratios to assist in achieving these objectives. However, the design exercise, whilst of use to the estimator and specifier, may not meet the needs of the concrete user, unless consideration is also given to:

(a) the amount and location of reinforcing steel;
(b) method of transport and distance to be transported;
(c) method of placing;
(d) the type of contractor and labour and plant available.

The ultimate test of any concrete is if it can be placed satisfactorily and the finished job is of the desired standard. Concrete is only well designed when it can be placed satisfactorily in a mould, form or deck.

The site agent will need to acquaint himself with design principles sufficiently well to be able to recognise any shortcomings that will prevent the obtaining of a satisfactory product.

Mixing

Whilst small amounts of concrete can be mixed by hand, mechanical mixing is desirable where uniform concrete is required. There is a wide variety of mixers and methods of filling available and the user needs to seek advice in selecting the most suitable equipment.

Batching and mixing plant

Batching is the process of collecting, storing and proportioning the constituents of concrete prior to mixing. Small mixers are

Fig. 2.1 General view of batching and mixing plant, with lorry about to receive load from mixer

loaded by hand using shovels; larger plants use mechanical loaders such as hand scrapers, scraper booms, bucket draglines or conventional dozers for hopper loading. Aggregates are generally weighed using load cells or lever arms, whilst the water is measured by a flow meter. Cement stored in bulk silos is also weighed by the use of load cells. For cold weather working the batching plant may also incorporate under-bin heating tubes and mixing-water heaters.

Batch mixers

Batch mixers are of four main types:

1. tilting drum;
2. non-tilting drum;
3. reversing drum;
4. forced action.

1. Tilting drum mixers are suitable for medium strength concretes in the capacity range 100 to 200 litres, and for producing mass concrete with large (150 mm) aggregate in 3 and 4 m^3 batch sizes. Materials are poured into the drum where the blades lift the concrete to the top of the drum and allow it to fall, encouraging mixing. Front to rear mixing is good provided the specified drum angle is maintained (generally 20° to 30° to the horizontal). To avoid clogging of the mixer drum the water inflow should be started first and the dry materials loaded into a wet drum. To prevent loss during charging the cement should be sandwiched between sand and aggregate in the weigh hopper. Material left sticking to the blades will impede the mixing action and affect the quality of concrete produced. Discharge is achieved by tilting the drum downwards.
2. Non-tilting drum mixers are very rarely seen nowadays. Materials are poured into the drum and discharged by a chute that pivots into the mixing drum, catching the material on its fall from the top. They are not suitable for high strength or lean mix concretes. Cycle times are about 3 minutes, the discharge being particularly time-consuming. The output capacity range for these machines is 200–750 litres.
3. Reversing drum mixers have two openings and two sets of blades. Materials are loaded at one side and efficiently

Fig. 2.2 Reversing drum. The weighing mechanism should be checked weekly to the full capacity of the scale and at all points between. Also, the machine must be checked regularly to ensure that ropes, wires and buckets are sound

mixed by one set of blades. When the rotation of the drum is reversed the arrangement of the second set of blades empties the concrete out of the other side. These machines are good all-round mixers and are found on some medium sized sites. They can cope with most concretes, except lean and sticky mixes and range between 200 and 500 litres output capacity.

All three types described above are known as free-fall mixers where use is made of gravity for mixing.

4. Forced action mixers only use mechanical power to mix the constituents. Consequently mixing is much more thorough and all types of concrete can be produced. They are most versatile but are most expensive in terms of energy consumption and mechanical wear. Capacities range from 200 litres to 3 m^3.

Pan mixers using forced action are the

(a) stationary pan with rotating blades on a central axis;
(b) rotating pan with rotating blades on an eccentric axis.

In both, the pan is mounted horizontally and discharge is through doors on the pan floor. Materials are usually weighed

Fig. 2.3 Aggregate storage being loaded from stock pile. Cement silo and 'pan' or forced action mixer are shown on the right. The mixer is bottom-opening and discharges direct into dumpers or lorries

whilst the previous batch is being mixed. With the fast discharge the complete cycle time will be between 1 and $1\frac{1}{2}$ minutes.

The other type of forced action mixer is the trough mixer which has paddles rotating about a horizontal axis set in a semi-circular trough.

Setting up and maintenance of mixing plant

To get the best value from any mixing equipment and to avoid breakdown, which can be very expensive, regular inspection is required. The following list of maintenance checks is of primary importance.

Checklist for mixers

1. Always follow the maker's instructions.
2. Keep clean – wash down at the end of a pour or day's work
   – hose out the drum
   – clean with a run of coarse aggregate and water for 10 minutes.
3. Keep the machine level for accuracy and less wear on moving parts.
4. Ensure the water gauge shows correct quantities.
5. Empty water tank in cold weather to avoid freezing.
6. Charge with stones and water during breaks in use.
7. Check blades regularly for wear and twisting.
8. Keep working parts greased.
9. Pay particular attention to engine maintenance and winch wires.
10. Ensure wheeled plant is rigidly fixed down.
11. Check that pneumatic tyres are jacked clear of the ground.
12. Check speed of rotation regularly by putting a chalk mark on the drum and counting the number of times it passes a given point in one minute; the operator is able to check if the machine complies with the manufacturer's recommendation.

Checklist for small independent weightbatchers

1. Check for level to ensure accuracy of dials.
2. Check scales against test weights.

3. Ensure buckets are free to pivot on knife edges and that load cells are not clogged with material.
4. Ensure knife edges are not worn.
5. Keep well greased.
6. Avoid impact loads (e.g. bags of cement).
7. Do not overload.

Checklist for built-in hydraulic weightbatchers

1. Ensure that the machine is level.
2. Check hydraulic system is not leaking oil.
3. Check scales against test weights.
4. Keep well greased.
5. Avoid impact loads.
6. Do not overload.
7. Keep underside of hopper free from accumulated materials.

Siting of mixer plants

One of the most important decisions that has to be made is the siting of the concrete mixing 'set-up'. To do this effectively, all available information on the items listed below must be considered:

(a) the overall job programme;
(b) the time period allocated to concrete – will some take place in winter?
(c) the total amount of concrete;
(d) plant available or could be used in the concreting process, i.e. cranes or hoists;
(e) plant already available within the company;
(f) the 'peak' or maximum pours and their frequency; the method of distribution, i.e. pump, crane, dumper, conveyors etc;
(g) power supply available;
(h) availability of materials and access required for delivery; services available (light, water, heat etc.);
(i) cost of alternatives.

The overall programme, together with the concreting period, will help in deciding the size of plant and whether to purchase or hire. It will also indicate if any special considerations need to be taken into account when siting the mixer for varying weather periods.

From the programme 'peak' pours can be assessed and from this information, together with the total concrete figures, the average daily concreting requirements can be determined.

These average daily pour figures will be useful when selecting the mixer capacity, providing the 'peak' requirements are also taken into account. For example, should the average pour be 50 m^3 per day with two or three pours during the month reaching 100 m^3, it could be more economical to equip the site with a mixer large enough to meet the 'average' demand and then short-term hire plant to meet the occasional 'peaks' rather than hire the larger expensive plant over the whole period. Alternatively, concrete needs may be supplemented by ready-mix or a second mixer.

Once the concrete has been mixed it has to be transported to where it is needed on site.

The various methods of conveying wet concrete are described in Chapter 6, but they need to be considered when deciding the mixer location in relation to site activities. The mixer needs to be positioned as close to that part of the site requiring the most concrete, the objective being to reduce the time and cost of transportation to a minimum.

The mixing of concrete which is repeated on site frequently, or over a long period, must be given close and careful consideration in relation to the planning of the whole programme. Work study techniques can often be employed to advantage and small savings in time can be significant when accumulated over the period of the contract.

An example of this would be the bulk supply vehicles travelling further on the site to reach the mixer facility. Such vehicles are larger and will be required less frequently so, providing good access is maintained, there is no extra cost in delivering sand, aggregate and cement to a point that reduces the distribution of the mixed concrete. Other considerations are power and water, but these can normally be conveyed above or under the ground so if extended to the most favourable mixer position, will only involve the one prime cost.

In residential areas, it is advisable to also consider the effect of noise and dust on neighbouring houses.

To enable the checking of both quality and quantity of the delivered aggregates and the quality of the mixed concrete, space and weather protection will be needed for the operatives if this function is to be carried out satisfactorily.

Summary

Seek advice in the careful selection of the most appropriate mixing and loading plant. Take care of the plant and carry out regular maintenance. Maintain good constant supervision checking the quantity and quality of the materials, the variations that always occur, and keep good records of tests and changing weather conditions.

Careful consideration of siting and drainage of mixer set-up will pay dividends.

The designer of a concrete mix must produce a workable, placeable concrete; charts and water/cement ratios are only half the story.

Suggested reading

1. Cement & Concrete Association. *Batching and Mixing Concrete on Site* (Man on the job leaflet).
2. Palmer, D. *Concrete Mixes for General Purposes.* Cement & Concrete Association, 1977.
3. Cement & Concrete Association. *Construction Guide: general purpose concrete mixes – a guide for builders.* 1979.
4. Cement & Concrete Association. *Concrete Practice.* 1979.
5. Illingworth, J. R. *Movement and Distribution of Concrete.* McGraw Hill, 1972.
6. Department of the environment. *Design of Normal Concrete Mixes.* HMSO, 1975.

3

Ready-mixed concrete

Introduction

The greater proportion of concrete used in the UK, about 60–70 per cent, is produced at ready-mixed plants located throughout the country. Ready-mixed concrete may be manufactured in two different types of plant:

1. Dry batching plants weigh the cement and aggregates then discharge them into a truck mixer. The water may either be added at the plant or on the site. In either case, the concrete is mixed in the truck.
2. With wet batching plants all the materials, including the water, are mixed in a mixer before being discharged into the truck, the truck only being used as an agitator.

The difference in techniques is important because concrete that has been previously mixed and then gently agitated does not need so much site mixing as the truck containing dry mixes requiring water to be added at site. The former needs 3 or 4 min and the dry mix 7 to 10 min.

Site co-operation

The production and delivery of concrete in a plastic condition has its own problems; for example, variable moisture content in aggregates, weighing vast quantities of materials, and traffic conditions.

The user requires assurance that each load is of a consistent workability and quality, yet these characteristics can change markedly between batching and placing.

Most ready-mix producers are able to produce concrete that satisfies these requirements and the majority agree to guarantee minimum standards of concrete for their customers.

The advantages to be gained from using pre-mixed concrete instead of site mixing are that:

(a) ready-mix plants can be sited on or near to sources of good aggregate. By specialising in the production of concrete, the ready-mix supplier accumulates considerable experience;
(b) specialist staff with sound technical knowledge can also be employed;
(c) materials, and consequently the concrete, will be more consistent and more suitable to the particular requirements of the site;
(d) site space is saved. On restricted sites ready-mix concrete is often the only way concrete can be supplied;
(e) cost control is easier;
(f) responsibility for quality control is transferred to the supplier;
(g) the experience of the supplier reduces the need for trial mixes; supplies are unlikely to be interrupted as the supplier will have adequate back-up resources;
(h) site labour can be reduced; concrete can be conveyed and discharged at different parts of the site, to save on distribution costs;
(i) it is cheaper and less inconvenient to reject unsuitable concrete.

The supplier has contracted to provide concrete to conform with the job specification or order. If it fails when properly tested it can be rejected at the supplier's cost.

Among the disadvantages of pre-mixed concrete are that:

(a) site roads and job access have to be constructed to carry heavy large vehicles;
(b) deliveries may be late due to traffic, breakdown of lorries or excessive demand at the works. These can all increase site costs;
(c) volume requirements have to be assessed more accurately and well in advance of delivery;
(d) small amounts of concrete, for example for a base or a few kickers, can be expensive. However, a small mixer on site will overcome this disadvantage;
(e) cancellations have to be made well in advance of delivery;

(f) it will take longer to stop manufacture and deliveries after a site delay;
(g) the longer delay caused by rejecting a load of concrete can have a serious effect on previously placed concrete.

To gain full advantage of the use of ready-mixed concrete the site will need to foster a good relationship with the supplier by maintaining good communications. Where large batches of concrete have to be handled and placed, careful, realistic planning is essential. Even on the smallest of jobs where ready-mixed concrete is to be used it will be advantageous for the supplier and the user to be in contact before the order is placed to discuss, where applicable, the following questions upon which the success or failure of the task may depend:

(a) which plant will be supplying the concrete? This could be important in an emergency, e.g. lorry breakdown;
(b) will the concrete be dry batched and truck mixed, or plant mixed? This could affect placing time and planning, as more time is required on site when using dry batched;
(c) will all the water be added at the plant or will some be added at site? This may affect the selection of site supervision as more knowledgeable staff will be needed to control the water;
(d) can the plant supply heated concrete? This could be important if winter working is envisaged;
(e) will the plant be reliable and have sufficient spare capacity in plant and trucks? The type and size of delivery truck will influence the type and size of site access;
(f) is agitation during transit permitted?
(g) what is the method of agreeing payment for delays (both parties) and part loads? Agreement will avoid discord before deliveries commence;
(h) what is the method of ordering, cancellation or postponement; by whom should it be made and how much notice is required?
(i) who will be authorised to sign for deliveries?
(j) what samples and/or trial mixes will be required?
(k) will changes be permitted after the original order has been placed and how will these be authorised and recorded?
(l) what will the completed volume be contained in each vehicle and the margin of error? This is important for day-to-day ordering and planning pours. If 1 metre is ordered

it should, after vibration fill a 1 metre container;

(m) is any reduction in stone content required for placing purposes? This is often required for deep lifts to prevent honeycombing at kicker level;

(n) how will doubtful loads be checked? There should be agreement as to the weighbridge to be used. Periodic checks should be made of loaded lorries.

Placing orders

Concrete required for a specific job will normally have details of the architect's or engineer's requirements contained in the job specification; all information concerning the manufacture of the concrete needs to be stated on the order. The type of concrete will be specified as follows:

Designed mix

Designed mix means that the contractor or manufacturer is responsible for selection of the mix proportions to achieve the required strength, but the engineer will be responsible for specifying the minimum cement content and any other properties to ensure durability.

Ordinary prescribed mix

Prescribed mix means that the engineer will specify the mix proportions from standard ones in BS 532 : 1981 to achieve the required performance; the supplier undertakes to mix these properly but responsibility remains with the engineer. The grade of a prescribed mix is a number which will normally (but not contractually) be its 28-day characteristic strength in N/mm^2.

Nominal mix

Typical examples of nominal mixes are 1.2.4. or 1.1½.3. These have largely been replaced, since they are insufficiently precise, by BS 5328 : 1981, recommendations for six basic ordinary prescribed mixes known as C7P, C10P, C15P, C20P, C25P and C30P. C stands for compressive strength, the number indicates crushing strength and 'P' indicates a prescribed mix, rather than a specially designed one.

A full list of tables and information are contained in the Cement and Concrete Association publication *Construction guide – general purpose concrete mixes – A guide for builders.*

Special prescribed mix

Special prescribed mix will be employed for example if a special aggregate or a different type of cement is required. The qualities of all materials used in this type of mix have to be specified.

Day-to-day ordering

Having discussed the main order requirements and classification, agreement should be reached on how these will be described when telephoning to avoid any possibility of the wrong mix being delivered.

The supplier should also be given as much time as possible – 24 hours is not unreasonable as a minimum if a reliable service is to be maintained.

He should also be given:

1. The name of the contractor.
2. The name and location of the site, the correct entrance if more than one exists, and the order number.
3. The concrete mix type, in simple precise and correct detail, including any special requirements particular to that day's pour; workability. This should reflect the placing conditions.
4. Other special requirements such as air entrainment, admixtures specified for a particular purpose or temperature of the concrete if cold weather is prevailing.
5. The total amount of *each* type of concrete required to be delivered.
6. The time to commence deliveries and the spacing of these to allow for site conditions and placing facilities. It is useful at this stage to make provision for procedures in the event of breakdowns or disruptions occurring at either end; the name of the dispatch clerk can also be useful.

Access

General access to the site should be provided to take a fully loaded truck weighing up to 24 tonnes. Since sites can change character overnight, a daily check on conditions should be made as a truck needs a turning circle of at least 15 m; this is particularly important – failure to check can create costly delays.

Checking deliveries

By giving due attention to checking deliveries the contractor can avoid subsequent time-consuming investigations, resulting in loss of goodwill between himself and the supplier.

The delivery ticket will contain:

(a) the name of the supplier and the depot;
(b) a serial number and the date;
(c) the number of the truck;
(d) the name of the contractor being supplied – name of the contract and its location;
(e) the grade of concrete;
(f) specified workability;
(g) type of cement;
(h) maximum size of aggregate;
(i) time of loading;
(j) quantity of concrete.

Special mixes will have additionally:

(a) cement content;
(b) type of aggregate if lightweight;
(c) type of admixture.

The contractor's representative who accepts the load is responsible for checking the ticket and must satisfy himself that the correct grade and quantity as ordered are being delivered.

When signing the ticket he should agree with the driver both time of arrival and completion of discharge.

Should any delay occur it is helpful if the site notifies the depot immediately, for they are probably relying on making further deliveries with the delayed vehicle.

Casual signing or lack of attention to detail at the time of delivery can prove very costly if the wrong concrete is discharged. Disputes are settled quicker and cheaper when supported by recorded evidence.

Further checks before placing

It is not enough to check only the paper work, it is also necessary to establish that the concrete meets the requirements of the order in regard to workability or testing.

Adding water

If the concrete has been mixed at the depot, or in a truck mixer, it should arrive on site with the ordered workability and therefore, no extra water need be added. If it is supplied from a dry batching plant, the delivery ticket will state the amount of water required and it is the driver's responsibility to add only that amount stated; no unauthorised person should be allowed to add water.

When the site agent asks and signs for additional water to make the concrete more workable, the supplier will no longer be responsible for the concrete meeting a strength requirement.

Testing workability

If checking for workability is not carried out properly (covered in more detail in Chapter 8), then disharmony and sometimes expensive litigation can arise. There is a need to maintain records which will vary to suit the site's requirements and conditions. British Standard BS 5328 : 1981 is applicable to both site-mixing and ready-mixed concrete and contains recommendations that should be adopted by the site. For example, to check the slump of concrete delivered in a truck it is now no longer necessary to take four portions at intervals from the complete load; this always presented the difficulty of having to place all the concrete before being able to check its compliance with the order.

Recommendation 9.2 in BS 5328 : 1981 states that a sample can be taken from the initial discharge after 0.3 m^3 has been discharged, as follows; 'The sample of approximately 20 kg mass shall be collected from the moving stream in a suitable container'. This sample is then remixed on a base plate and sub-divided into two specimens; each specimen shall then be tested in accordance with BS 1881 and the average of the two slumps shall be the slump for compliance purposes.

Discharging the ready-mix concrete

When the user is satisfied the concrete is as ordered, the truck should be discharged as rapidly and efficiently as possible.

A fully laden truck can be emptied in about 10 minutes, but this is not always practical. The supplier normally expects the truck to arrive on site and complete its discharge within 30 minutes. It is in everyone's interest that this is achieved but it needs careful pre-planning.

It is preferable to be able to discharge the concrete directly from the truck to where it is required. This means that a check on the access is necessary prior to arrival. If the site is large or complex someone should be detailed to pilot the concrete to its discharge point.

Fig. 3.1 Discharging ready-mixed concrete

At the discharge point the gang, together with the working vibrators or any other plant, should be ready and men fully informed of their tasks.

A truck has a maximum discharge height of about 1.5 m and by using a chute extension can cover a semi-circular radius of about 3 m from the back of the truck.

In addition to checking the condition of the access road, it is very important to look for overhead cables that may be fouled by the truck and also to check on the safety factors at the pour location. For example, the shoring in trenches may not be sufficient to take the weight of the loaded truck.

Finally, the truck needs convenient wash-out facilities which should be provided at a location on the site that will not cause a disruption to deliveries or nuisance to the general organisation of the site.

Summary

Fuel and labour costs have made ready-mixed concrete expensive; to obtain full advantage of its use the site agent needs to:

(a) carefully plan the site to receive large heavy trucks;
(b) seek whole hearted co-operation with the supplier by ordering correctly, in time, by being efficient in checking and discharging and by notifying the supplier of any undue delays;
(c) discuss programmes, large pours and site conditions with the supplier;
(d) carefully estimate the quantity required;
(e) carry out tests in the approved manner and carefully record any unusual facts;
(f) be well aware of the latest methods of testing deliveries and fully instruct the workforce staff.

Suggested reading

1. Cement & Concrete Association. *Ready-mixed Concrete* (Man on the job leaflet).
2. Cement & Concrete Association. *Concrete practice.* 1979.
3. British Ready-Mixed Concrete Association. *Code for ready-mixed concrete. Specification. Ordering. Production.* 1975.

4

Formwork

The accepted definition of formwork, formerly called 'shuttering', is: 'a temporary structure built to contain fresh concrete so as to form it to the required shape and dimensions and to support it until it hardens sufficiently to become self supporting'. Formwork includes the surface in contact with the concrete and all the necessary supporting structure. The term 'falsework' is used to describe any temporary supports to an engineering structure during construction, such as bridges, tunnels, dams and formwork.

Formwork can be regarded as a concrete on-cost; consequently, the cheapest, most readily available and easily disposable materials are used.

Only where repetition, consistent quality and safety are major factors is it necessary to consider using the more expensive durable materials such as timber, steel and plastics. The final appearance of the structure very much depends on the quality and workmanship of the formwork and, because this can account for up to 65 per cent of the total cost of the structure, it is a very important item.

Formwork design factors

Multi-storey construction allows the repetitive use of formwork which is designed to obtain maximum re-use, simplicity of erection and rapid striking.

Other factors to be considered in designing formwork are:

(a) improved hoisting facilities now available;
(b) the wide variety of materials available;
(c) the demand for high quality and decorative finishes;
(d) increasing labour and material costs;
(e) the trend towards larger pours and larger structures;
(f) safety.

Improved hoisting facilities

Prior to the general use of cranes on sites, formwork units were restricted to a size manageable by operatives. Metal forms were generally 0.625 m × 0.625 m, and plywood forms or decking were seldom larger than 2.4 m × 1.2 m.

Currently, the tower crane allows the use of much larger units, although there are limitations in use as outlined under site considerations.

Variety of materials

Materials are discussed in detail later in this chapter. Suffice it to say at this point that the designer and user of formwork must have knowledge of all the currently available materials to ensure that the most suitable and economic are selected.

Fig. 4.1 Prefabricated formwork for a housing scheme – metal strongbacks

Demand for high quality and decorative finishes

In the past the framework of a building was commonly clad with facing materials such as stone, marble and brickwork, but now intensive efforts are being made to develop decorative finishes for structural concrete. For these reasons formwork designers and users now aim for much higher quality formwork and a greater consistency of performance in order to provide large external elevations in 'acceptable', decorative and durable patterning (see Chapter 14).

Labour and materials costs

The increasing cost of materials, combined with the effects of a limited skilled labour force, had necessitated much research into the planning and design of formwork. For instance, a sharp rise in timber prices caused designers to reassess the use of timber in formwork in an endeavour to employ the material more economically or to find an alternative.

Trend towards larger pours and larger structures

The trend towards larger concrete pours and larger structures has come from increased technical knowledge, the greater capacity of concrete mixing and handling equipment and the demands made by the size of many modern projects. As a result, formwork design has moved away from the site to the engineer's office where pressures, stresses and material strengths can be calculated to permit these large pours to be carried out in safety and to the exacting demands of the design team.

Safety

The hoisting of large panels while work proceeds and the general emphasis on site safety have also meant that reliance can no longer be placed on the foreman carpenter to solve all formwork requirements. Lifting tackle has to be adequate and since loading, the weather and siting for cleaning, maintenance and storage need detailed consideration, specialists in formwork design are essential. This does not mean that the skills of the man on the site can be discarded, for the good formwork designer knows how to employ those skills to the best advantage.

Selection of a formwork method

Depending on the size or complexity of the job, most, if not all, of the following factors will affect the final choice:

(a) quantity of formwork required and contract period;
(b) removability and striking;
(c) re-use;
(d) consistent surface finish;
(e) consistent strength;
(f) fixings required;
(g) specification;
(h) method of hoisting, storage and protection;
(i) availability of materials;
(j) labour availability.

Quantity of formwork required and contract period

In planning formwork it is better to assess the work as columns, beams and slabs rather than linear metres or m^2, for example.

The period allowed for erecting formwork and the time of year may also influence both quantity of formwork and the type of material selected.

Removability and striking

Specifications can place restrictions on striking times that will affect the early removal of forms. Forms may have to remain undisturbed until the concrete reaches a minimum strength, until it is sufficiently cured, of the required colour, or to protect it.

Re-use

When considering re-use attention has first to be given to repetition, then conversion and finally the residual value for future use.

Repetition is of fundamental importance and early discussions with the architect and engineer should seek to rationalise sizes towards this aim.

Casting larger or longer columns first or boxing out of large forms, then converting forms to the smaller sizes is an example of economy in formwork material.

The quantity and quality of the formwork can be increased to advantage if its future use can be forecast for another contract. This will have the effect of speeding up job A, and jobs B and C, etc will be more economical if forms can be stored or transferred to these following contracts.

Consistent surface finish and strength

These will be related to specification, quality and durability of selected material and number of re-uses.

These factors are interlinked when the formwork material is chosen.

It is important to establish just what the specifier expects from his detailed description and it has to be considered that at its final use the form should still be in a condition to produce the desired finish. Therefore, the quality, durability and number of re-uses are related in providing the specified finish throughout its use.

Fixings

The majority of concrete structures need holes and chases for the concealment and distribution of services and for fixings for windows and doors. The choice of forms will depend on the size and frequency of these provisions.

Should screws, nails or bolts be necessary to fasten the timber forms, the re-use value will be reduced. A high quality or metal form is not required or suitable in this case.

Specification requirements

Since it is difficult to accurately describe a particular finish or standard, architects and engineers resort to the use of loose terms such as 'fair face' or 'blemish free'.

The contractor will be well advised to provide, at the earliest opportunity, a large sample, preferably also incorporating a joint, to represent his interpretation of the specification. It should be to a standard that can be reasonably maintained throughout the contract. If this standard is agreed early in the construction stages the task of monitoring quality by both the client's representatives and the site supervisor should be easier.

Methods of hoisting, storage and protection

It is sensible to make the forms as large as possible to utilise any mechanical plant available on site, thereby reducing site labour. There must, however, be adequate storage space for the protection and cleaning of forms between uses.

Storage space should be considered carefully to minimise the transporting between uses. For example, temporary platforms on the external scaffolding of high-rise buildings would save lowering and raising formwork from ground level as work progresses.

Availability of materials

The availability of certain materials in some areas may be limited so before a final decision is reached a full investigation should be made. Reject materials can sometimes be obtained that will be suitable for the temporary job that is required from some formwork.

Labour availability

Another factor affecting the choice of formwork is the quantity and quality of skilled labour available. Should there be a shortage of skilled labour, consideration should be given to using ready-to-assemble formwork (proprietary formwork).

Proprietary formwork is available in standard sizes to meet a wide variety of uses. It can be assembled easily to a regular pattern which simplifies checking and supervision of both sizes and the fixings. The forms are robust and are not easily damaged by a lack of skill in handling and transporting.

Even in areas where skilled labour is available it is advisable to check for any preferences of method. Success is more likely if local labour is familiar with the selected method.

Materials for formwork

Timber

Timber is the most commonly used material, not only due to its inherent favourable characteristics, but because carpenters usually erect formwork they naturally prefer to work with this material.

The basic advantages of timber are:

(a) ease of working and handling;
(b) adaptability for fixings and tolerances;
(c) durability, particularly of new plywoods;
(d) easier striking and repair due to the flexible nature of timbers and plywoods;
(e) good insulation properties;
(f) the possibility of removing damaged sections and re-using the remaining portion, of re-facing or altering for other formwork or other site uses.

It is important to remember that timber not only shrinks and swells but can also twist out of shape. Care must, therefore, be taken in storage and re-use. Shrinkage and distortion have, however, largely been eliminated in modern plywoods by combining them with plastics and resins. These new plywoods are waterproof and have very high re-usable values.

In comparison with steel, timber provides better protection against frost in view of its superior insulating properties.

Fig. 4.2 Timber formwork for a high quality pre-cast product

Cardboards and hardboards

Cardboard cylinders give a good finish for 'one-off' columns and can be obtained in a variety of sizes. They can, if required, be left in position to act as protection throughout the contract.

Tempered hardboard of external quality provides a good finish for a limited number of uses if care is taken to avoid buckling by providing for the slight movement that occurs when the board absorbs some moisture from the concrete. It should be glued to framing rather than nailed to prevent buckling.

Hardboard is also useful as an inexpensive stop end, being easy to scribe around, or over, projecting steel reinforcement, then suitably stiffened against concrete pressures.

Steel

Where considerable re-use is required steel formwork is now being employed on an increasing scale. It is particularly suitable for the construction of multi-storey buildings and large civil engineering projects.

For precasters and makers of repetitive items of concrete, steel is an ideal formwork material.

Advantages of steel formwork include:

(a) consistency of surface quality and size of unit;
(b) less likelihood of failing under load;
(c) simplicity of erection and supervision;
(d) the possibility of casting larger pours without bolting;
(e) robustness giving the ability to withstand rough site handling;
(f) the availability to hire for planned periods, thereby enabling costs to be more accurately assessed.

Disadvantages are:

(a) the pattern left on the concrete is not always acceptable and not easily varied;
(b) if purchased, a long-term evaluation on its use must be made to estimate the return on capital;
(c) blowholes are more likely;
(d) the smooth surface finish obtained can cause problems when applying other finishes;
(e) heat losses are greater, therefore frost damage is more likely;
(f) fixings are not so easy to apply. Drilling damages the forms, which are costly to repair.

Proprietary formwork

Proprietary systems comprise panels, either all metal, or plywood with metal frames, together with a variety of telescopic centres, adjustable props, strong backs and ties to support or hold the forms together. They can be used for floors, walls, columns and beams, etc. where repetition, simplicity and speed are required.

Large pours, more exacting tolerances and the need for less labour to fix bolts and cut timber have contributed to the increasing use of proprietary panels and steel strong backs, struts and proprietary ironmongery. An additional bonus offered by the suppliers of proprietary equipment is an excellent advisory and design service backed by experts having many years of specialist experience.

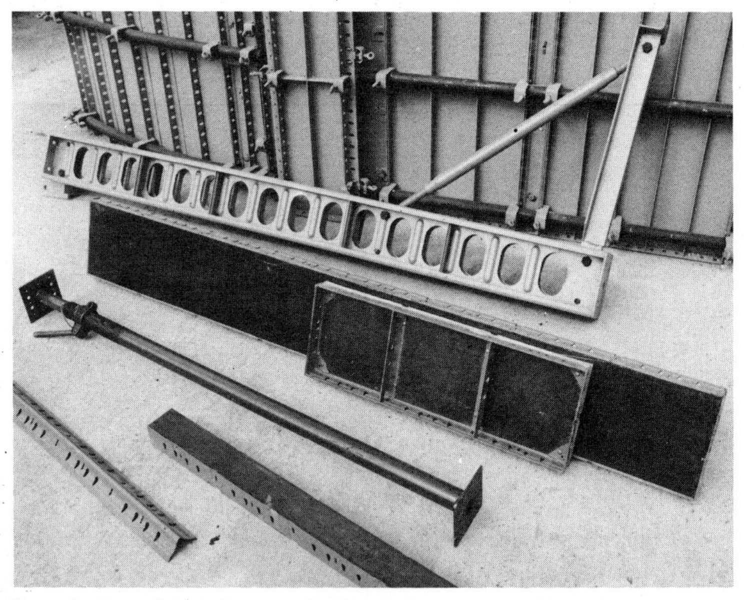

Fig. 4.3 Proprietary formwork. From the top: assembled proprietary formwork showing standard tubular scaffolding to align and strengthen panels; strong back used as heavy duty (upright) soldier to reduce bolting through concrete on deep forms; panels that are obtainable in a wide range of sizes; adjustable prop; make-up panel; corner piece

Permanent formwork

Permanent formwork is that which is left in position after pouring the concrete. It sometimes assists in carrying the load as part of the structure, but usually it is merely a convenient way of forming concrete. For example, plastic sheeting or old bricks can be used to reduce the overdigging of foundations and to contain concrete and these materials are not recovered. Slim prestressed concrete planks are also used to span traffic areas, reducing the amount of support work below and thereby permitting access whilst supporting the concrete above.

New materials are always being sought and one that is gaining attention is glass reinforced cement (grc).

Fine strands of special glass, able to resist the chemical action of the cement, are used together with cement to produce a thin, very strong material. The glass provides the tensile strength which enables this material to take the weight or pressure of the concrete without having to be supported from below.

Site considerations in the selection of formwork

The site's influence on the selection of approval of a formwork system will be based on:

(a) the experience of the site manager and his staff;
(b) the size of the formwork task;
(c) time or programme allowance;
(d) availability of labour;
(e) hoisting or horizontal movement facilities available;
(f) access on and around site;
(g) type and amount of scaffolding;
(h) refurbishing and storage areas;
(i) striking times. These vary with the weather and must be considered in relation to the period of the year when the largest work load occurs;
(j) number of uses for each different form;
(k) residual value or convertability of forms;
(l) type of concrete and method of filling forms.

The foregoing consideration will lead to a choice of one of the following:

(a) manufacture of formwork on site;
(b) employment of an off site specialist to make the formwork;
(c) hire or purchase of proprietary equipment;

(d) use of sub-contractors;
(e) site or off-site precasting.

Manufacture of formwork on site

There are advantages to be gained when choosing to make the formwork on site, always provided there is sufficient room for a covered working area and labour available at a time when preliminary work of setting out, hutting, etc. has also to be tackled. When site management and personnel are involved with the manufacture of the formwork and its subsequent use, they will have some familiarity with and acceptance of the methods selected. The forms can also be made to the quantities required as the work progresses.

Off-site manufacture

The big advantage of off-site production is that manufacture can proceed independently of the pressures of the site. Special skills are made available by employing specialists, so that delivery dates and costing can be more accurately forecast and controlled.

Site use of proprietary systems

The effective and economical use of system formwork is very much dependent on good site planning, organisation and possession of full information. The majority of users hire equipment, only purchasing if a large number of repetitive jobs are anticipated. Hiring gives the possibility of concreting in large pours.

By hiring for a specific job the costing can be more accurate but this does depend on the site completing hire within the forecast planned period. Delays can be expensive and the care and return of the forms also becomes a cost factor which can easily get out of hand on a disorganised site.

Use of sub-contractors for formwork erection

By employing specialist sub-contractors the main contractor aims to boost his skilled labour force and, having negotiated a price, hopes to have a better idea of the final cost. There are several ways of employing additional skilled groups; for example labour only, labour and materials, or the complete package of formwork, steel and concrete. The success of any of these methods depends on good planning to ensure continuity. Also needed is a very clear contractual agreement on those areas where disagreement frequently occurs, such as the amount of formwork to be employed and the fixing of the small items like pockets, stopends, features and other labour items. These are not always obvious and not always included on the drawings.

Site or off-site pre-casting

Big savings in formwork costs can be made by early and detailed analysis of the structural drawings.

Repetitive elements such as staircases, balconies, beams, copings, etc. can be pre-cast either on or off the site. By this method, a robust mould, for example a repeat staircase, can be used repeatedly to cast the flights, saving both time and money. In addition the pre-cast staircase will provide access during construction much earlier.

If space or site experience is lacking there are many suppliers who can produce units under factory conditions to reduce site formwork and to provide a consistent, standard quality product.

Erection

Due to the temporary nature of formwork the importance of the pressures and loads that are applied when concreting are sometimes overlooked.

It is the task of site management to make sure all necessary fixings, bolts, props, etc. are in position and able to remain secure whilst the forms are being filled. A check list is useful if prepared to suit the particular type of formwork and site. It could include the following:

1. Are the forms being used in their correct place and sequence; are their identification markings right and plain to see?
2. Are props, walings, bearers, clamps, ties in the right quantity and at the spacing shown, or agreed, on drawings or sketches?
3. Are props vertical, braced if required and on a firm base? (*Note*: The base material can vary from day to day. For

example, ground could have been frozen or wet when erection started.)

4. Have grout checks been placed in position at joints to prevent leakage of grout?
5. Are all fixings, boxings out, fillets and pour height indicators in position and firmly fixed, allowing for later striking?
6. Has adequate access and guard rail–toe board protection been provided to enable the concrete gang to work efficiently and safely?
7. Have any sloping or horizontal forms been checked? These are subjected to upward pressures and need to be bolted down or otherwise restrained.
8. Have large sections been checked having been marked with their weights, to ensure that these weights are within the crane's working radius? Are lifting devices capable of carrying this weight?
9. If unfamiliar formwork is used do all operatives understand the maker's instructions? Have they any required special tools, as lack of knowledge or improvisation can cause accidents.
10. Are all temporary distance pieces removed, forms free of sawdust, etc. and wash-out pockets replaced?
11. Are all props sound and placed upright, with the correct pins and spreaders if ground is doubtful? Are props fastened at base and head?

Action prior to concreting

A routine procedure should be arranged and followed on all concreting jobs, depending on their size and complexity, to check the forms before concreting. This check, in addition to the supervision of the erection processes should cover the following items:

(a) the correct number and type of ties used and in the designed location;
(b) the securing of the stop ends;
(c) alignment and levels;
(d) the ability to strike the formwork without damage to the concrete. (*Note*: Bent splice bars are easy to insert when the form is empty but they make it difficult to remove the mould or stop end when the concrete has hardened);

(e) the release agent applied;
(f) the reinforcement, its cover and spacers;
(g) the removal of all the tie wire loose ends. These will cause staining on the concrete face if left in the bottom of the mould. (Note: A magnet on a length of tying wire is a useful method of removing these from closely reinforced beams ready for concrete);
(h) safe and sufficient access for placing and compaction;
(i) availability and condition of the necessary equipment vibrators, lighting, hand tools, skips and chutes, covers and heating in cold weather all need checking over and reserves made available to cover breakdowns during concreting.

Concreting

During concreting there should be continuous attendance to watch the formwork and to deal with any signs which indicate that a dangerous situation may be developing.

It has been known for wedges, props and bolts to become eccentrically loaded or to work loose, needing just a pause in filling for adjustment to prevent disaster.

Tell-tale devices such as plumb lines or gauge rods should be fixed appropriately so that a continuous check can be maintained on line, level or any other movement.

Formwork failures

Despite all precautions, failures can still occur due to a faulty bolt or a loose wedge or strut not being noticed.

When this occurs whilst filling a beam, wall or column failure usually commences with a grout leak, followed by a bulge in the formwork.

The immediate action is to stop concreting. Where a tie fails some of the remaining ties become overloaded and progressive failure can occur. Consequently, pressure has to be relieved and controlled.

It is dangerous to attempt to employ extra propping, strutting or cramps in an effort to force the formwork back into position. Serious accidents have resulted from this as the whole assembly can easily be pushed over.

Having stopped concreting the wet concrete should be removed to below the bulge. In narrow walls and beams this is not possible and the choice is either to take off one form to clear the concrete and then reassemble, or to wait sufficiently long for the concrete to harden enough to stay in position whilst one form is removed. It is then safe and relatively easy to cut away the bulging concrete beyond the steel, then replace the form, followed by the concrete, as expeditiously as possible. The replaced concrete will bond to the clean, recently poured remaining concrete.

Any remedial work should, if possible, always be agreed with the controlling engineer before being carried out. Should action be necessary in his absence then he must be notified as soon as possible of the steps taken to effect the remedy. More detailed methods of repairs to concrete are dealt with in Chapter 13.

Stop ends and day joints

Concrete construction is a continuous process but it needs to be temporarily stopped for a variety of reasons.

Joints needed for structural purposes, for example expansion joints, and which are included in the design are called movement joints.

Joints made necessary by interruptions which are not part of the design are called day joints. The reasons for these are:

(a) the limited amount of concrete that can be mixed and placed or delivered in any one work period;
(b) the manpower, formwork or finishing resources available.
(c) the economical use of formwork;
(d) breakdown, weather, etc;

Whatever the reasons, joints need to be formed accurately and positioned with care, attention being given to the following:

(a) correct positioning and support to projecting starter bars or continuity reinforcement;
(b) grout tightness and easy removal;
(c) placement in the area of least reinforcement;
(d) agreement of the consulting engineer with regard to positioning; careful removal of stop ends as soon as the concrete has hardened sufficiently to stay in position. Consideration should be given to the alternative of forming permanent stop ends with expanded metal or precast concrete.

Checking after filling forms

Forms and supports are sometimes disturbed during concrete placing. Consequently, immediately on completion of columns and walls, they should be checked for plumb and line while the concrete is still plastic. The outside of the forms should be cleared of any concrete or grout since it is easily removed at this stage but not after it has hardened. It can also make striking more difficult.

Any distance pieces used to hold wall panels apart can now be removed if not taken out as the filling proceeded. If unsleeved bars or bolts are being used to form holes or hold formwork together, they will need to be turned or eased before the concrete sets and then removed as soon as the concrete is hard. Forms left overnight are more difficult to remove resulting in damage to forms or concrete.

Stop ends to slabs and walls can also be removed as soon as concrete hardens enough to support itself. These stop ends are invariably notched around projecting continuity steel and are, therefore, difficult to remove the longer the concrete is allowed to harden.

Striking

Striking times vary according to:

(a) weather conditions (mainly temperature):
(b) type, size, shape and position of the element in the structure;
(c) guidance given in the job specification.

Most vertical forms, e.g. columns and wall forms, can be struck the day after concreting but the concrete will still be green and easily damaged unless care is taken. In cold weather it may be necessary to delay striking. Colour variations can also occur when forms are removed early.

Suspended or load-carrying forms for beams and slabs will need to stay in position while the concrete gains sufficient strength to be self supporting – normally 11–15 days in summer. This time can be reduced if propping is designed to allow the majority of support and formwork to be removed whilst a minimum amount of propping remains in position. The practice of striking all the formwork and then re-propping the slab or beam is not to be recommended because it is not easy to assess if the member is under or over loaded by the replacement props.

A double-headed prop (Fig. 4.4) enables the majority of the formwork supports and panels to be removed for early reuse. The second head remains in position to provide limited support to the slab while it continues to gain strength.

Specified striking times should be strictly observed. The specifier may require, for example, curing before stressing or additional loads may have to be added to the immature part of the structure.

Bolts, ties or cramps should all be loosened gradually to prevent the last tie from binding and so damaging the concrete. All bolts and loose parts should be collected in a box or basket to prevent losses. Where forms need prising away from the concrete, a gently tapered timber wedge should be used. Nail bars will damage both concrete and forms.

Any fillets forming drips and boxing out should be left until they have dried out and shrunk, thereby facilitating removal.

Soffitt formwork also needs care in removal, starting from the support ends and working towards the beam or slab centre; this will avoid overloading and possible deflection under its own load. 'Crash striking' of large areas of formwork, where large areas are dropped in one go, must not be allowed. It is dangerous and can cause damage, not only to the formwork but also to the structure by sudden loading.

Lifting positions must be clearly indicated and other trades kept away during striking and removing. A tail rope is useful to control movement when lowering panels in windy conditions. Care must be taken to prevent damage from projecting scaffold.

Cleaning, maintenance and storage

Provision must be made for the removal and storage of large sections of formwork. A level storage area is required and space necessary to permit cleaning and refurbishing of large forms.

The sooner the forms are cleaned the easier the task. On timber forms only a stiff brush will be needed but if grout has been allowed to become hard and stubborn then a hardwood scraper should be used, not metal scrapers.

GRP and other plastics only need rubbing over with a wet cloth for cleaning but both timber and plastic should be treated with a release agent after cleaning. Metal panels will need a light coating of oil before storage to prevent rusting.

All formwork needs to be carefully stacked and stored. Panels

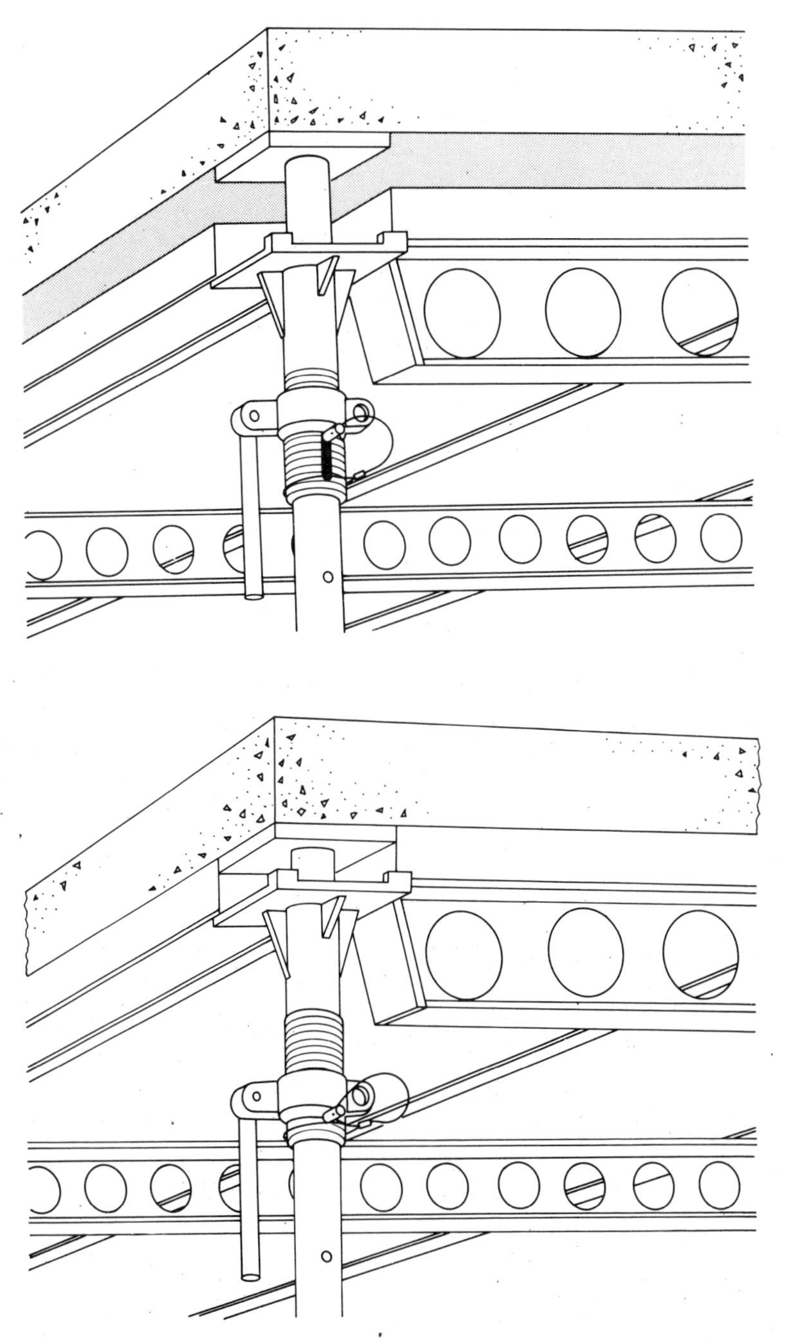

Fig. 4.4 Double-headed or 'quickstrip' adjustable props

and column forms are best kept horizontal and face to face. The forms and components should be clearly marked and kept together for easy identification on re-use. A tidy store reduces wastage, damage and losses.

Release agents

The main function of a release agent is to make it easier to strike the form but it also helps to preserve forms constructed from timber.

There is a wide variety available for specialist purposes for application to particular surfaces.

On sites where a number of different formwork materials are used it is safer to use a chemical release agent, as this can be applied to timber, steel or plastics.

Should a special release agent be specified, the container must be clearly marked and the agent applied to the maker's instructions.

Release agents should be applied with care as too much will stain the concrete. A spray is the best applicator, but a brush or roller can be used.

New timber forms absorb release agents so they should be given a coat of the appropriate release agent at least 36 hours before using and then a second coat immediately before fixing.

Summary – formwork

Formwork has a very important role in concrete construction, since most concrete has to be formed or contained in some shape. Efficiency and economy can be achieved by following the recommendations made in this chapter.

To be successful, the user needs to:

(a) be conversant with the characteristics of the many materials used and their availability;
(b) always give consideration to striking, removal and the re-use of mould materials;
(c) be able to adapt and improvise whatever resources are available and plan their use, having regard to any limitations of manpower or skills;
(d) take full advantage of the specialist knowledge that is often freely available in both commercial and educational fields.

Suggested reading

1. Richardson, J. G. *Formwork Notebook* (2nd edn) Viewpoint Publications, 1982
2. Cement & Concrete Association. *Formwork* (Man on the job leaflet).
3. Baker, E. M. Formwork and site management, in *The Practice of Site Management*, Vol. 1 (2nd edn). Chartered Institute of Building
4. Austin, C. K. *Formwork to Concrete* (3rd edn). George Godwin, 1978.
5. Richardson, J. G. *Practical Formwork and Mould Construction* (2nd edn). Applied Science Publishers, 1976.
6. Cement & Concrete Association. *Concrete Practice.* 1979.

5

Steel reinforcement

Steel and concrete have similar coefficients of expansion which allows steel to be used in structures where concrete alone would be inadequate. Concrete has a very good compressive strength which enables it to withstand heavy direct loading – for example, in mass concrete bases – but in tension or bending situations, such as a slab or beam, the very small concrete strength is ignored and steel is included to take these loads. Steel can do this very well, providing it is placed in the correct position and is protected by the surrounding concrete from the effects of the weather.

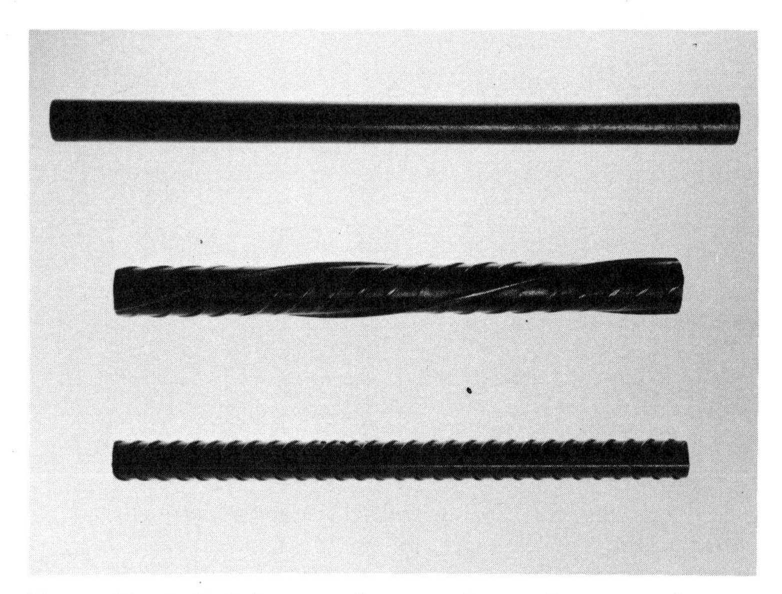

Fig. 5.1 Types of reinforcement in present day use. From top to bottom: round mild steel; hot rolled high yield bar; cold worked ribbed high yield bar

Successful concrete construction depends very much on sound knowledge of reinforcement.

It is important to be aware of the different types of reinforcing steel, how to identify them and where to use them.

In schedules 'R' for Round is mild steel in the form of plain smooth round bars.

'T' denotes steel which has a ribbed appearance and may be twisted (cold worked). This used to be known as high tensile but is now called high yield steel.

It may be one of the wide variety of specialist steels. It is, therefore, necessary to look for an explanation of any abbreviations given in the specification or drawings.

In order to avoid some of the difficulties that occur where several types of steel are on site it is advisable to:

(a) keep the different types in separate areas or compartments;
(b) arrange, and have generally understood, clear marking either by coloured paint on the ends of firmly fixed durable labels on each different type;
(c) when in any doubt seek advice.

Care of steel

Careful selection of the storage area for steel can pay good dividends. Items needing special attention are as follows.

Cleanliness

It is difficult to keep steel clean, so although it is desirable to have the store accessible to the delivery lorry and lifting equipment, steel needs to be stacked away from the route to the canteen or general areas – chestnut fencing or similar will prevent site personnel from taking short cuts across the steel.

Weed killer, blinding concrete, ashes, etc. can be used to prevent vegetation from growing amongst the steel.

Concrete strips or timber sleepers should be used to keep the steel straight, off the ground and away from the rising damp.

The supplier needs to be notified of the maximum lifting facilities available so that steel can be bundled to suit – splitting bundles of steel on arrival in order to unload can be very costly.

Methods of supply

Nearly all steel is cut and bent in factories prior to being delivered to site. A small bending machine may be found on site to take care of last-minute alterations to the design. Some large civil engineering sites have their own cutting and bending facility.

Steel bars may be obtained direct from the steel mills, but early notification and sufficient quantities of each type are needed to make this economical. Or they may be bought from stockists, who carry a variety of steel types and sizes, making it simpler to order a mixture of types for the medium or smaller contracts. Bars are normally 12 m long.

Companies supplying cut-bent and labelled steel are often also able to provide a fixing service.

Site cutting and bending

A reinforcement designer invariably provides what is known as a bending schedule. This is prepared in a standard manner and gives details of the type, quantity and location of all bars; it is also customary to give the total length of the bar and details of bending by reference to a BS shape code number.

A useful exercise before commencing cutting is to rearrange the listed order, which is often dictated by location, and to put the bars in length order, then, by first cutting the longest bars, waste can be reduced.

To carry out cutting and bending efficiently as much room as possible should be made available. Twelve metre bars may have to be bent at each end and it may be more convenient to have a bending machine at each end of the bench to avoid the inconvenience of turning a long bar end to end. Provision should be made for stacking short ends upright so that these can be sorted easily.

Correct practice in bending

Heat must not be used for bending high yield steel. Large diameter mild steel bars can be heated to a dull red colour for bending but must not be cooled in water.

A hand or powered bending machine should be used which has been regularly serviced and equipped with the correct size mandrels.

The bench must support the bar properly as it is being bent to avoid bending in the wrong plane. Waste can be avoided before bending a large section complex bar if a template is made with a small section bar for checking. Gauges should also be provided for the checking of steel diameter on delivery, as it is difficult to identify the different sizes by observation. The current British Standard (BS 4466 : 1981) – 'Bending dimensions and scheduling of reinforcement for concrete' should be used to obtain an acceptable standard of bending.

This specification gives maximum bending parameters for both mild and high yield bars. This information is used to produce bending schedules. A cutting tolerance of approximately +25 mm should be allowed. As this can be lost in bending when a bar has to be bent both ends, marking should commence by measuring away from the centre of a bar. Any surplus is then taken up in the hook, leaving the bar correct in those dimensions which are important. All bending machines vary in performance, so if a bar has several bends and some critical dimensions are to be maintained, then a trial bar should be used to set up the correct parameters on the machines.

Regular maintenance is essential as is the replacement of points of greatest wear. Steel loses some flexibility in cold weather and care must be taken in bending and handling at temperatures below 5 °C.

Fixing

Having stored, cut and bent the steel in reasonable conditions the task is now to transport and fix it. Before fixing it needs to be clean. Slight rusting does no harm but any loose flaky rust and other surface material such as paint, mud, mould or oil needs to be removed. When moving over rough ground some support may be necessary to prevent long bars from distorting. Correct slings should be available to hoist steel without excessive bending.

To ensure that bars stay in the correct position during concreting they are tied at the intersections. Once assembled these cages, as they are called, should not be exposed for too long, as rust can mark formwork or previously poured concrete.

Tying is normally carried out by using 16–18 gauge soft iron wire or patent fixing clips, care being taken to select the appropriate sizes. Loose ends of the tie must either be cut off or bent so that they cannot do any harm by rusting and showing through the concrete face.

Time can be saved by prefabricating stock repeat items such as beams and columns, taking care to mark these clearly for later identification. If facilities are available, spot welding of links instead of tying with wire makes the reinforcement cages safer and easier to transport.

Steel cover

Cover is the word used to describe the amount of concrete needed by the designer between the bar and the concrete face. It is important that this is correct at the completion of concreting.

Table 5.1 gives nominal cover of dense natural aggregate which

Table 5.1 Nominal cover to reinforcement

Condition of exposure	*Nominal cover*				
	Concrete grade				
	(mm)				
	20	*25*	*30*	*40*	*50 and over*
Mild: e.g. completely protected against weather, or aggressive conditions, except for brief period of exposure to normal weather conditions during construction	25	20	15	15	15
Moderate: e.g. sheltered from severe rain and against freezing while saturated with water. Buried concrete and concrete continuously under water.	—	40	30	25	20
Severe: e.g. exposed to driving rain, alternate wetting and drying and to freezing while wet. Subject to heavy condensation or corrosive fumes.	—	50	40	30	25
Very severe: e.g. exposed to sea water or moorland water and with abrasion	—	—	—	60	50
Subject to salt used for de-icing	—	—	50*	40*	25

* Only applicable if the concrete has entrained air.

should be provided to all reinforcement, including links, when using the indicated grade of concrete under particular conditions of exposure, but in addition it may be necessary to specify concrete mix details to provide the required durability. For example, an exposed beam of a lifeboat station would require concrete of grade 40 or more with a cover of 60 mm (very severe condition).

A similar size beam in a 'mild' condition of exposure, such as the internal beam in an office block, could have any grade of concrete from 20 to 50. If 30 was chosen to protect the steel from fire then a cover of 15 mm would be adequate.

Where bars are required in two directions, the cover specified refers to the bar nearest to the surface, as shown in Fig. 5.2

The following is an extract from the BS Code of Practice for the Structural use of concrete: CP 110: Part 1: November 1972. Clause 3.11.1.4.

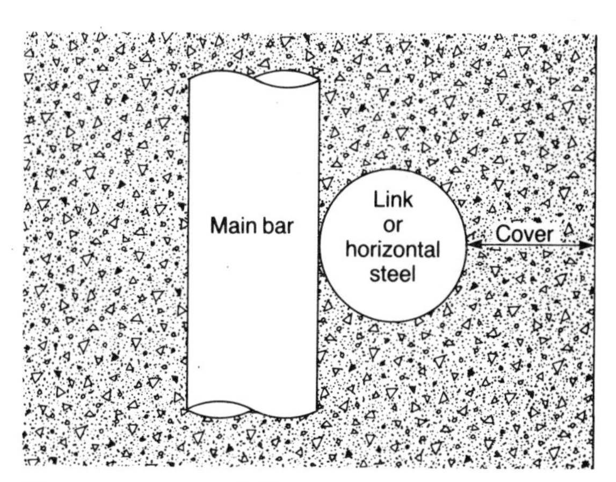

Fig. 5.2 Concrete cover to reinforcement

> Cover to reinforcement should be determined by considerations of fire resistance and durability under the envisaged conditions of exposure.
>
> The nominal cover should always be at least equal to the size of the bar and in the case of bundles of three or more bars should be equal to the size of a single bar of equivalent area.
>
> Where a surface treatment such as bush hammering cuts into the face of the concrete, the expected depth of treatment should be added to the nominal cover.
>
> Where a member due to its particular situation is required to resist the action of fire for a specific period, the normal cover

may need to be increased or, alternatively, the concrete cover to the main bars needs to be reinforced to prevent premature spalling.

Special care should be exercised for conditions of extreme exposure or where lightweight or porous aggregates are used.

Spacers are used to hold the reinforcement away from the forms and in correct position. They are made from a variety of materials – concrete, plastic and asbestos being the most commonly used. Spacers are designed to stay in position and permit concrete to be placed and compacted all round. They often need to have sufficient strength to support not only the steel but also the concrete and the men placing it.

Fig. 5.3 Plastic and concrete spacers

Concrete spacer blocks are sometimes made on site by using 2 : 1 sand and cement mortar and planting in some fixing wire. This wire must be kept back away from the surface to prevent subsequent rusting. Alternatively, the wire should be galvanised. It is bad practice to use brick, wood or stone as spacers. It is of equal importance to arrange a system of inspection or provide templates to ensure that any splicing or continuity steel left projecting from the concrete on completion of the day's filling has not been pushed out of position during the concreting. Starter bars should be inspected and reinstated if necessary.

If starter bars are likely to be exposed for longer than a couple of weeks they should be covered with plastic sleeves or similar protection to prevent rust staining. Care must also be taken to ensure that bars projecting through stop ends or forms remain parallel to each other. Bars bent in different directions may result in the formwork having to be destroyed on striking.

The top mat or mesh to a floor slab is often supported from the bottom mat by 'chairs'. These are proprietary items or may be bent from mild steel rod to the desired height and need to be of sufficient strength to support foot traffic whilst concrete is placed.

Fabric (or mesh) reinforcement

Steel fabric is often used in floors since it is preferable to the tying of loose bars. It is most suitable when the plan is not complicated. Too many openings or complicated shapes result in offcuts of mesh which are not easily re-usable.

Welded fabric is usually made up of main bars with lighter bars arranged at right angles, spacing the main bars correctly in accordance with the designer's requirements. Drawings will indicate the location of the mesh by T for top and B for bottom or by the direction of a diagonal line. The direction of the main bars will also be shown.

As various strengths and spacings are used in different positions, the amount of lap must also be shown on drawings and strictly applied on site. Drawings will indicate the outline of the fabric and the direction of the main bars and so should be followed very carefully.

Fabric is generally indicated by heavy dotted lines on drawings and by a BS reference number. Care must be taken to stack different types separately with a painted board marker fixed close to the stack. A label on the top sheet of mesh will soon disappear if only part of the stack is used.

Where cover is critical, flat sheets are preferred to rolls of mesh. Chairs and other supports necessary for mesh should be detailed on the schedules.

Top mesh can be supported by packing or hanging from tubular screeding rails. This mesh is frequently light in character and has to be positioned near the top surface of the concrete to prevent surface cracking.

If spacers are not permitted on a vertical wall face, timber packings can be used to keep the steel in the correct position, providing these are arranged so that they can be withdrawn as the concrete proceeds and checked after the pour to be sure all have been recovered.

Splicing reinforcing bars

The method of providing a continuous length of steel by lapping one bar past another or by the use of a mechanical splice should only be carried out where shown on the drawings or with permission of the engineer in charge.

Blinding concrete

Blinding concrete is a thin layer of concrete, approximately 75 mm thick, positioned in the bottom of foundation bases or trenches or on ground floor formation. This enables the bottom steel to be placed correctly and also assists in keeping the steel clean during fixing.

Information

Before commencing any fixing, a check should be made to be sure the latest drawing is being used – old schedules and drawings must be discarded as soon as revisions are available.

Reinforcement for prestressed concrete

Steel tendon to be used for prestressed concrete requires special care and attention. It must not be rusty, pitted or contain kinks.

It must be stored in a damp-proof, ventilated shed or store and the bundles stacked clear of damp ground.

Inspection should continue throughout the process but with skilled knowledge and attention.

Summary – reinforcement

Be aware of the different types of steel used for reinforcement. Understand bending schedules and the basic design reasons for the positioning of reinforcement.

Order reinforcement steel in the construction sequence.

Store it so that it is available in the construction sequence – moving steel is costly.

Choose a method of supply to suit the site – limited space for storage often results in the nearest bar being cut and the remainder wasted.

Have the right appliances for cutting and bending, and maintain them.

Prefabrication can save overall site time.

Site staff will need to be acquainted with steel fixing drawings and to be able to check these before concreting commences.

Suggested reading

1. Cement & Concrete Association. *Plastic bar spacers*. 1972.
2. Cement & Concrete Association. *Concrete practice*. 1979.

6

Transporting concrete

Virtually all concrete has to be transported from where it is mixed to its final position. It can be a simple job of discharging it down a chute from a ready-mix truck into a trench – on the other hand it may have to be conveyed long distances or pumped to great heights.

The incorrect choice of transporting or placing equipment can easily cancel out the care, thought and cost that have been given to produce a workable concrete of the correct strength and durability – a skip that leaks or a lorry that is dirty can seriously reduce the qualities of good concrete.

Concreting is a process that commences with the decision to use concrete as a structural material and is only completed when the unit is put to use and satisfies its desired structural, aesthetic and durability requirements.

To fulfil these requirements, consideration at the mix design stage must be given to the method of placement and the type of formwork and its location.

Decisions on both mix design and methods of transporting sometimes have to be made before the contract has been awarded. In this case site management has to face a compromise on mix design and placing methods. If, for example, an average workable mix is specified to be transported by a dumper over rough ground for some distance, it will most likely arrive in a segregated condition, which is not satisfactory.

Should it not be possible or economic to provide suitable access or to change the method of transportation, then the remedy is to adapt the mix design to suit the conditions by providing a more cohesive mix with means of compacting this once it is placed in the forms.

The objective is to have the concrete conveyed and placed so

Fig. 6.1 Ready-mixed concrete discharging directly into trench

that it is in a condition to surround the reinforcement properly and completely fill the form.

Selection of transportation method

The best solution for a particular project is usually reached after consulting company personnel who can provide information and advice on the following:

(a) access available to and on the site;
(b) plant availability (company's own hire or purchase from plant companies;
(c) quantity of concrete (peak demands and periods);

(d) quality of concrete (specification);
(e) material (its availability – consistency and specification requirements);
(f) quality and number of staff available.

The route from a ready-mix plant can pass through crowded city streets causing deliveries to be unpredictable; site mixing may therefore be the better solution. Alternatively, a limited amount of site mixing could be provided as an additional assurance for continuity of concrete placing.

The resources of the site must be sufficient. There must be the right type of plant, space to use it, sufficient power available and men with the necessary skill.

Not only should the gross amount of concrete that has to be obtained and placed within the contract period be determined, information will be needed on the peak demand period and on the frequency of large pours.

The specification needs to be studied to establish the type of concrete that will be required, its appearance and durability; each requirement having a possible influence on the choice of transport.

The availability of the specified materials and the way these will be delivered must be investigated. A check should also be made of the staff and labour not only in regard to numbers but also in the range of skills available. To select a concrete pump or tower crane without having a skilled operator would be foolhardy.

The foreman carpenter must be aware of the rate of pour and the equipment to be used for placing and compacting. As concrete pumped into a column exerts considerably more pressure on the forms than placing by barrow and bucket, the formwork has to be capable of taking this additional pressure.

Similarly, the ganger can often influence the final decisions on this vital matter. It is frequently overlooked that of all the people involved in the concreting process, the ganger probably sees more 'wet' concrete than any other and ought to be able to assist by making known his requirements and preferences for scaffolding, storage space on access platforms, chutes, tremmies and other plant. Too often one sees discarded on sites a complicated piece of equipment, 'purpose made' to solve a placing problem, which has proved too heavy or otherwise inconvenient in practice, mainly because the designer has not been aware of all the problems to be solved. He should consult site staff as appropriate to prevent this wastage.

Methods and plant

Manual handling

The escalating cost of manpower means a hard look must be taken at all forms of concrete transport involving labour. Nevertheless, there are jobs where wheelbarrows, prams or motorised prams are most suitable and have the advantage of low plant hire cost and offer alternative employment to men when they are not concreting. A man will be noticed if standing idle, but costs are not always so obvious when expensive pumps or cranes are not fully employed.

Dumpers

Dumpers are the logical progression from the motorised pram for the horizontal wheeled movement of concrete. They come in a wide variety, ranging from the common forward tipper to types having hydraulic tipping and turntable discharge action. Choice is dependent on the particular application. Roads or other means of access need to be reasonably level to avoid segregation of the concrete – heavy aggregate tends to travel to the bottom of the skip if conveyed on uneven ground for long distances. Mixes may have to be redesigned to give a more cohesive quality if the dumper has to traverse rough, uneven ground.

Dumper drivers and dumpers can prove expensive if not fully employed on productive work. Once a decision is made to use this method, concreting or alternative work should be programmed to obtain full use of man and machine.

This can be accomplished by either increasing the amount of formwork together with an increase of concrete output in order to relate this to the selected method of transportation.

The alternative is to choose a machine that can also be used for conveying other building materials.

Care must be taken to avoid contamination if using a dumper for carting 'muck' or other purposes; washing-out facilities must be provided convenient to the supply of concrete.

Lorry-mounted concrete transporters

On larger contracts much use is made of lorry-mounted transporters which can carry 2 m^3 or more. These often have re-mix blades and tipping facilities operated by a hydraulic ram. When fitted with extension chutes large amounts of concrete can be transported and placed directly into the forms. When mounted on a four-wheeled drive chassis bad ground conditions can be overcome and the need for temporary access reduced.

Lorries

Larger amounts of concrete for roads, dry-lean mixes and mass concrete fills can be tackled by side- or rear-tipping lorries. These are used when fairly long hauls are required. It is essential to provide protection to the wet concrete against the effects of wind, rain and sun and this is accomplished fairly simply by fitting a tarpaulin on rails above the lorry. Washing facilities are also very necessary if trucks are used for other materials.

Fig. 6.2 Concrete being discharged from rear tipping lorry, Note, top right of illustrations: tarpaulin folded back, available to cover concrete during transport to protect it from wind, rain or sun

Ready-mix trucks

Ready-mix trucks are a very convenient way of transporting concrete, and the concrete should be discharged if possible, directly into its final position.

To enable the contractor and supplier to get the best value from ready-mixed concrete the following points should be noted:

- (a) order full loads whenever possible;
- (b) give plenty of notice;
- (c) provide full information on specification, noting any special details such as air entrainment, etc;
- (d) describe fully the location and condition of the site (a fully laden lorry can weight 20 tonnes);
- (e) state the size of the job, starting date and approximate daily and peak requirements.

It should be appreciated that significant savings can be achieved if good access, good organisation and communications with the plant operator, informed supervision and control can be provided.

Vertical transportation

As the lifting of concrete adds considerably to the final cost, careful consideration needs to be given to the final choice of method.

The amount and frequency of the pours will dictate whether plant, e.g. a static or mobile pump, is installed for the exclusive movement of concrete, or whether plant, e.g. a hoist or crane, is to be employed which will carry out other functions in addition to moving concrete.

Small portable hoists of the winch type can be employed for minor pours, such as lifting a loaded barrow on low-rise houses.

Larger hoists can be fed directly by a site mixer or ready-mix truck and can be arranged to tip automatically at the height required into a storage or floor hopper. Concrete can then be distributed by any of the 'horizontal' methods previously mentioned.

The use of a storage hopper divorces the upper distribution of concrete from the hoist/mixer cycle and so speeds up the concreting process.

Tower or mobile cranes

A careful look must be given to the unit costs before deciding on craneage for concreting. It is seldom good economics to use a tower crane exclusively for concreting. However, there are many situations when craneage is essential for other purposes and in these circumstances the spare craneage time can be used for concreting.

When choosing craneage it is also necessary to have knowledge of or to seek advice on, the type of skip; these are now manufactured in such a wide variety that it is nearly always possible to obtain one most economical and convenient for the desired purpose.

Skips

Skips and buckets for use with tower cranes are a common method of combining vertical and horizontal movement of concrete.

'Skips' are those containers having a capacity of below 1 m^3 and 'buckets' is a term used for the larger containers, although there is no clear dividing line when using these descriptions.

There are three main types of skip:

1. roll-over skips;
2. concrete-pouring skips;
3. dual-flow skips.

Roll-over skips are designed to lie flat for filling, so reducing their height to allow direct discharge from a standard mixer.

Fig. 6.3 Concrete dual-purpose pouring skips. Front and bottom. discharge with light hinge tray – lever or handwheel gear operation. The tray is moved only for required pour position

When hoisted they travel in the upright position on the crane hook and a lever-operated or geared wheel door assists easy controlled discharging into formwork.

Concrete-pouring skips stand upright and can be obtained with either side or bottom opening and are designed to take the full discharge of the common range of mixers.

Lever-operated discharge permits either partial or full discharge; the side discharge model is particularly useful for filling columns or walls. Dual-flow skips are versatile in having a low overall height that enables them to be filled under most mixers. A movable chute allows either bottom or side discharge, so these skips are useful on smaller sites having a variety of needs.

Costs

The cost of a crane and operator can equal the cost of employing 8–10 men and this is a useful yardstick; it follows that good supervision is required in order to employ this piece of equipment fully. Where several trades have a demand on craneage time and priorities, a man at general foreman level is needed to allocate and direct crane usage. The site manager needs to seek expert advice to enable him to choose the most suitable crane from the extremely wide variety available.

Concrete pumping

Mobile and static pumps

The 150 mm static pump with metal quick-release pipes has been available for quite a while and is still suitable for large pours spread over considerable distances. It has the advantage that a pipeline can be taken over undulating ground without the preparation and access needed for wheeled vehicles.

Lorry-mounted boom pumps

The lorry-mounted boom pump – usually with a 100 mm discharge – has rapidly increased the interest in concrete pumping and consequently has improved the technique and performance of these versatile pieces of plant. So much so, that there is a temptation to employ pumps exclusively, even where other methods would be more appropriate and economic.

As with all other methods of moving concrete it is necessary to evaluate the economic factors. There is no magic figure of 50 m^3 per day that decides the economics of pumping; a couple

Fig. 6.4 Concrete being placed at first floor level by means of a static pump. Also shows vibrator being used and timber formwork

of cubic metres for a base behind a busy occupied shop or office may well justify taking a flexible pump line through the directors' dining room, yet a 100 m^3 pour may be tackled by using agitating trucks and corrugated iron chutes.

The advantage of pumping is the ability to move concrete both vertically and horizontally at the same time. Pumps can transport concrete more than 60 m vertically or 300 m horizontally and can discharge 30 m^3 to 100 m^3 per hour.

The performance in height, distance and discharge depends on the pump type, the horizontal and vertical length of the line, the number of bends in the pipe and the type of concrete. Also to be considered is the amount of concrete the gang can receive, handle and compact properly and the rate that the concrete can be delivered.

It follows that concrete pumping needs careful planning and co-operation between the contractor, the concrete supplier and the pump hire company who should meet well in advance of the proposed pumping operation to agree the following:

(a) type of pump;
(b) concrete mix specification;
(c) items to be concreted, their access with pipelines and sequence;
(d) total quantity and required rate of delivery;
(e) pump locations together with access for truck mixers;
(f) maximum distances, horizontal and vertical, to be pumped;
(g) washing-down facilities.

The concrete supplier is responsible for ensuring that the mix will be suitable for pumping over the proposed distances and for meeting delivery requirements.

The contractor is normally responsible for providing:

(a) suitable and safe access to the place where the pump will be placed to work;
(b) enough piped water at this working position and washing-down facilities;
(c) cement for the initial 'paste lining', or grouting, of the pumpline. About 50 kg is needed for each 20 m length of pipe;
(d) any extra labour needed for pipe erection and dismantling;
(e) suitable and adequate supports and/or anchorages for pipelines;
(f) the concrete supplier and pump line company with information regarding delays or breakdowns.

Site organisation

Efficient use of site pumping makes considerable demands on the site manager and requires forward planning. Attention to the following points will help to make the operation successful:

1. Easy access needs to be provided for the mobile pump and there must be enough room to allow the ready-mix trucks to turn around and back up to the pump hopper. This

Fig. 6.5 Two mobile concrete pumps discharging into basement formwork

space needs to accommodate two trucks at the same time so that as one finishes its discharge the second one can commence. This will keep up the necessary continuous supply of concrete to the pump.

2. The site of both pumps and ready-mix trucks should be reasonably level and firm and needs to be situated so that

pipelines are as short and straight as possible.

3. A plentiful supply of water will be needed for cleaning and flushing out the pump. Therefore, the area must be suitably drained. Washing out of trucks should preferably be carried out in a separate area.

Since formwork will be filled rapidly it should be designed to have enough strength to meet the pressure imposed. More supervision may be needed to watch for early signs of possible failures.

The pumping rate and the delivery rate have to be matched as well as the rate at which the concrete can be placed and effectively compacted and finished. Increases in manpower may well be required to meet the concrete outputs that can be achieved.

Pumpable concrete

Concrete for pumping has first to meet the requirements of the specification but it also has to be pumpable. The design expertise needed to achieve this is beyond the scope of this book. Advice is readily available from specialist pumping contractors or ready-mix suppliers.

However, the site manager needs to be aware of the designer's objectives to enable him to identify the source of any faults in the pumping process. The pumping mix must not be prone to segregation or bleeding and needs to have a low frictional resistance to enable it to be pushed along the pipeline.

This can generally be obtained by having,

(a) a target slump of 75 mm;
(b) a cement content of at least 280 kg/m^3 to ensure complete filling of the voids in the combined aggregates;
(c) a uniform aggregate grading with no gaps and a minimum of voids;
(d) a slightly increased sand content over that normally used – an additional 50 to 75 kg/m^3.

Aggregates to avoid are flaky or crushed materials.

Safety in pumping

Very high pressures may be used when pumping concrete and this can be extremely dangerous. These operations must, therefore, not be carried out carelessly and all concerned need to be well briefed in the operation of the process and the safety precautions.

Pneumatic placers

With pneumatic placement, concrete is conveyed along a pipeline by first feeding it into a container which is then sealed. By using compressed air the entire contents – usually the output capacity of the mixer – are forced along the pipeline. The process is then repeated.

By emptying the pipeline for each mix it is claimed that the likelihood of blockages is reduced. However, sharp bends should be avoided, particularly near the blowing chamber. Workability and cohesion of the concrete are very important; advice from the manufacturer is essential and should be followed closely.

The device is particularly suitable where space and access are limited or difficult, for example mining and tunnelling. Particular care needs to be taken at the discharge or 'wet end' by shielding from the effect of pressure at the completion of each batch.

Chutes

Chutes could be utilized more, as a considerable proportion of concrete is lowered into basements, footings, etc. They need to be portable to enable them to be easily dismantled for cleaning at the end of a pour or day, as washing them in position may ruin the completed concrete. Disused semi-circular metal column formwork is excellent for this purpose. Hire of the extensions or used ready-mix discharge chutes can be negotiated with the supplier.

Corrugated iron fixed to a light timber frame using a stiff broom to assist the flow of concrete can be a simple, cheap alternative.

Portable self-drive conveyors

Portable self-drive conveyors can be hired at reasonable rates from most plant companies and are economical in manpower. They can be used by themselves, in conjunction with chutes for gentle slopes, or for lowering concrete, and can be arranged in series to permit concrete to be conveyed up or around obstacles. They need to be quickly removable and portable, otherwise washing down can cause damage if left in their position of use.

The latest model is easily portable and combines conveyor and chute principles. It has a towing bar and pneumatic tyres and so

Fig. 6.6 Concrete being conveyed directly from mixing plant to road slab

provides a useful extension to truck mixers. It incorporates a water tank and pump for washing-on completion. Placing capacity of 30–35 m^3 hour and belt speeds up to 120 m/minute are claimed.

First floor slabs, roofs to single-storey factories, columns and walls can be filled quickly and efficiently.

Elephant trunking and PVC tubing

For deep and inaccessible pours, concrete can be kept cohesive by using elephant trunking or PVC tubing. The latter can be cut off or raised as the pour progresses. Elephant trunking comprises short, circular, metal tapering tubes that can be suspended from each other by hooks and chains allowing the discharge end to be removed as the pour progresses in height. They can be usefully employed for columns or walls.

Cement paste losses

In nearly all the techniques used in the distribution of concrete, allowance must be made for the initial loss of cement paste when starting work.

It is good practice to reduce the coarse aggregate in the first mix, in order to coat the mixer interior and thereby prevent the first mix out being 'lean'; this practice has to be extended throughout the whole process. The chute, conveyor, tremie, or skip will all retain a certain amount of paste for which an allowance must be made. This practice must also be extended to the place of discharge.

Some grout or paste will 'hang up' on the formwork and steel, so the first mixes need to be 'fatter', i.e. contain more paste to avoid 'honeycomb' defects at the base of a column or wall.

Care and maintenance

To achieve reliable performance, in particular from concrete transporting plant, reliable servicing is essential.

After use, all skips, pipelines, etc. need a thorough washing to clear all cement paste and concrete spillings. A light coat of diesel oil or release agent will prevent concrete from sticking to the plant and assist when cleaning. Continuous attention to cleanliness and maintenance will be rewarded by a smooth running site.

All mechanical items such as gate-operating devices need oiling and checks made on oil and fuel intakes.

Summary

Quality concrete will suffer and site efficiency will be reduced if the equipment for transporting concrete is chosen carelessly or without fully considering the many factors that affect the most suitable method.

The main objective is to get the concrete to the point of placement as quickly and cheaply as possible and in the best possible condition.

The variables are many; the nature of the site, the size of the job, the type and condition of access, loading and discharge heights, size of aggregate and the concrete workability, are some of the factors to be added to choice of plant and their performances.

On some jobs the answer may be a combination of methods. When, for example, the concrete needs to be moved both vertically and horizontally a hoist with automatic discharging skips may feed a crane or dumper for final placing.

Only observation and experience will help the site manager to make the best decision.

Suggested reading

1. Cement & Concrete Association. *Placing and compacting concrete.* (Man on the job leaflet).
2. Jury, W. A. Concrete pumping, in *The practice of site management* (Vol. 1, 2nd edn). CIOB.
3. Baker, E. M. Transporting and placing of concrete, in *The practice of site management.* (Vol. 1, 2nd edn). CIOB.
4. Illingworth, J. R. *Movement and distribution of concrete.* McGraw-Hill, 1972.
5. British Concrete Pumping Association. *The manual of pumped concrete.* 1978.
6. Building Research Station. *Guide to concrete pumping.* HMSO, 1976.
7. Cement & Concrete Association. *Transporting and pumping concrete.* (Man on the job leaflet).
8. Cement & Concrete Association. *Concrete practice.* 1979.

7

Placing and compaction

Having given careful attention to the selection of materials and plant to produce good concrete, all can be wasted unless placing and compaction are given equal consideration. Concrete placed and/or compacted carelessly will result in a poor quality final product.

A number of factors have to be taken into account before deciding how to place the concrete, and when and how to compact it.

Before commencing to place concrete in its final position, a check should be made to ensure that the insides of the forms are clean and have been treated with a release agent. Any 'wash out' pockets that were provided in the bottom of the forms to assist cleaning must now be firmly re-fixed. The steel should be inspected to see if it is correct and fixed as shown in the drawings and also has sufficient 'spacers' to ensure that it stays in its correct position whilst the concrete is placed.

To enable the concrete to arrive at the placing area in good condition precautions must be taken to prevent exposure either to heavy rain or, equally damaging, drying winds or hot weather. Moving the concrete quickly and/or a simple cover on rails will greatly assist in avoiding the effect of rain, wind and sun.

Wind-blown dust or dirty containers can also cause problems when placing concrete.

Placing concrete

The main objective when placing concrete is to deposit the concrete as close as possible to its final position, quickly and efficiently to minimise segregation.

Specifications often incorporate time restrictions on the placing of concrete once it has been mixed. These have to be adhered to unless the clause is qualified before the contract has been signed.

Code of Practice CP110 suggests that concrete subject to delay can be used and that no harm will ensue provided that the concrete can still be placed and effectively compacted without the addition of further water.

Inspectors and site agents should look at the concrete rather than their watches. Temperatures vary as do concrete mixes, so strict adherence to a time limit can be non-productive.

No water should be added to the concrete once it has left the mixer, otherwise the properties might be adversely affected.

The concrete should not be allowed to fall in heaps and then moved along the form. It should be spread evenly, to ensure placement as near as possible to its final position.

The concrete gang must be instructed on how the concrete will be placed and compacted.

The most suitable plant should be provided, for example a crane skip with a controlled chute if filling walls with baffle boards to direct the concrete to prevent spillage over the forms.

Lighting will be needed inside narrow wall or similar formwork, as it is essential to be able to see the concrete as it is being placed.

There is a trend towards deep lifts, columns and walls with large civil engineering projects because time is saved by reducing the number of horizontal joints.

The placing of concrete successfully in deep lifts requires the concrete mix to be designed to minimise the risk of segregation or bleeding. The concrete should also be discharged down some form of trunking as this reduces the likelihood of damage to the forms and misplacement of the reinforcement. Trunking will also prevent the loss of grout which, if allowed to remain on the forms and steel above the concrete level, will result in 'hungry' concrete at the bottom of the pour. The hose of a concrete pump acts in the same manner as trunking, providing it is placed at the bottom of the pour at commencement and slowly raised to suit the filling process.

Should trunking not be available then the first mix of concrete to be placed should have extra 'grout or paste' as described previously in the chapter on methods of transporting.

Having gone to the trouble to select and grade and carefully mix together aggregates, cement and water, every endeavour must be made to avoid segregation when placing the concrete.

Segregation can occur in transporting mixes of high workability where the vibration of the lorry, dumper, etc. will cause the mortar (cement paste) to flow away from the coarse aggregate.

With low workability concrete, the larger aggregate can segregate from the bulk of the concrete.

Mixes low in sand content are more likely to segregate than cohesive mixes. Segregation results in varying appearance and quality of the hardened concrete.

Prevention will mean altering mix proportions or changing the method of handling or transporting the concrete. Concrete that has segregated slightly in transit may be improved by turning it over with a shovel.

When concreting with a skip it should be moved along the length of the pour to avoid discharging the entire skip at one point, as this would probably cause segregation and also abnormally load the formwork.

Long walls having a height of 2–3 metres will undoubtedly have through bolts. If using one poker head it will have to be continually withdrawn to the top of the formwork in order to vibrate in layers. In these circumstances additional pokers should be employed. They could be operated by one person, depending on the rate of pour.

Vibration is a means of compaction and should not be used as a method of placing by causing the concrete to flow. Segregation would probably occur and show on the completed work.

Vibration

Why is it necessary to vibrate concrete? The water in concrete does two things: first, it initiates a chemical action with the cement which unites the aggregates to form a hardened material. Second, it allows easy placement of the concrete.

Only about one half of the amount of water is required for the chemical reaction, the remaining half assists in the placement of the concrete.

The amount of water is the most variable factor in concrete. The greater the amount, the easier it is to place the concrete. But if there is too much water this will then weaken the concrete, too little water and the concrete is honeycombed. A balance needs to be struck between strength and workability.

Should the aim be ease of placing, i.e. workability and strength of secondary importance, then concrete with a slump of 75 mm plus would be satisfactory and could be compacted by hand tamping. The need to employ mechanical means of compaction would depend more on the complexity of the mould or the amount of reinforcement. On the other hand, where strength is essential and particularly where accessibility is difficult – for example, placing concrete in a form containing stressing anchors plus their surrounding reinforcement – these factors require a mix designed with a low water/cement ratio, and almost certainly would need mechanical means of compaction, usually an internal vibrator or poker.

Internal vibration

An internal vibrator or poker is essentially a vibrating tube at the end of a long flexible drive. It is available in sizes ranging from 25 mm to 75 mm. These need to be used and handled with care; the flexible drive should not be used twisted or with a sharp bend, it must not be left idling in any one position for very long in the concrete, nor should it be used either to drag concrete along or to spread heaps of concrete.

A poker vibrator should be selected so that it will pass easily through or alongside the reinforcement. The concrete is then placed evenly in layers 500 mm deep; the poker is then inserted vertically and quickly, being held in position just long enough for all air bubbles to cease coming to the surface. It should then be slowly withdrawn, allowing the concrete to flow into the space vacated. This process is repeated at not greater than 0.5 m intervals. Vibration can be regarded as completed when the surface of the concrete glistens. This will vary according to the workability of the concrete, the size of the section being formed, the rate of pour, the type of vibrator and its efficiency.

The depth of the concrete layer should never exceed 600 mm and the poker should always be lowered to at least 100 mm into the layer below. The main purpose of vibration is to expel the occluded air by compacting the component materials and closely interlocking them. Each 1 per cent of air remaining in the concrete reduces its strength by 5 per cent.

There will be some spots where the poker cannot reach but this can be overcome by combining the poker with external vibration.

A Kango hammer fitted with a rubber vibrating head will serve this purpose if it is used on the external formwork and follows the progress of the pour. The alternative is to have a higher workability concrete in these areas with the attendant danger of affecting the colour. In this case more cement must be added to offset the effect of the higher water content. When a high lift is reaching completion it is advisable to have the final layer of concrete with a slightly lower workability. This will take up the accumulated laitance (slurry) which has resulted from the vibration of the lower layers.

This final layer of concrete, because it does not have the pressure of the overlying concrete, will need additional compacting to remove air pockets. This can be done by using external vibration or by pushing a flat spade or similar tool up and down the inside face of the formwork. This is called 'slicing'.

On occasions when either appearance or structural requirements are critical, the concrete in high lifts of walls or columns can be carried on a few millimetres past a previously fixed fillet which marks the desired height. This additional concrete of doubtful value can be removed while the concrete is 'green' (usually a few hours later) to leave a very good joint surface.

Fig. 7.1 Poker vibrator in use

External vibrators

External vibrators are occasionally used but they are restricted to heavy robust forms that are used for repeated manufacture of, for example, large deep beams having webs which are inaccessible with a poker vibrator.

These vibrators are fastened to the outside of the formwork, which has to be designed to withstand the stresses and vibration created.

Vibrating tamping beams can be used to compact slabs of about 150 mm. Deeper slabs should be vibrated with a poker and only finished with the vibrating tamper. These beams combine the operations of screeding with compacting.

Vibrating tables are used in precast manufacture for repeat items such as kerbs, bollards, etc. Moulds are filled with concrete and placed on the vibrating table which vibrates and effectively compacts the product.

Overvibration and undervibration are loosely used terms that need some explanation. As previously stated, the purpose of mechanical vibration is to produce strong concrete from a mix with a low water content. Therefore, if the mix is correctly designed

Fig. 7.2 Twin vibrating beam

there is very little danger of overvibration. Should the concrete be separated by the poker vibrator and pools of grout form around the poker while it throws off the aggregate, then in these circumstances the poker is probably unnecessary as the mix is plastic enough to need only gentle pushing with a blunt batten of timber. This up and down motion is called punning.

There is a greater danger to concrete of 'undervibration' since this leads to honeycombing, which causes considerable loss of strength.

Care of vibrators

All vibrators must be treated with care and be properly maintained if breakdowns are to be avoided.

Poker vibrators are prone to failure if not looked after properly; the eccentric mechanism that provides the vibration sometimes jams in the head. A gentle bang will sometimes set it going again, but this method is not a 'cure-all'. When an important pour is being prepared it is wise to have a reserve vibrator either on site or readily available.

Site management must have and use the manufacturer's instruction booklet to follow the recommendations for both operation and maintenance.

Some additional points of care and maintenance are:

(a) check the voltage and frequency of electrically powered vibrators before connection to any power supply: check the equipment has a good earth connection;

(b) check periodically that petrol or diesel engines are running at the speed recommended by the manufacturer. If incorrect the frequency developed in the poker head will not be correct either and the compaction of the concrete will not be as quick and efficient as it should be;

(c) check all equipment regularly for signs of wear and ensure that all faults are remedied;

(d) check the grease in the bearings: if insufficient the vibrator tube may twist and jump about. When this happens, stop the vibrator, examine the bearings and grease if necessary;

(e) the poker should not be left in the same place for long periods when concreting;

(f) the poker should not be left running while waiting for fresh supplies of concrete;

(g) when using a pneumatically-driven vibrator, always clear the air line of moisture before coupling it up;
(h) when storing air-driven vibrators hang them to drain away any moisture.

A poker vibrator and its power connections are cumbersome tools which, when operated properly, need frequent movement to follow the layering of concrete placing. To achieve this, scaffolding should be robust and free of obstructions while concreting a long beam or wall.

Care must be taken to avoid disturbance to any fixings to the formwork. They should be firmly fixed with screws in preference to nails and very clearly marked on the outside of the forms to assist the poker operator to avoid displacing them. At the top of a beam or wall it is more difficult to clear the accumulated air that has been brought up with the vibrator. To clear this and any cracks that have been caused by the settlement of the concrete around the reinforcement, it is permissible to re-vibrate the top 450 mm about half an hour later.

Summary of placing and compaction

This important function can only be performed satisfactorily by paying care and attention to the following:

(a) reduce the time and distance of concrete movement to the minimum;
(b) select the most suitable and most reliable plant and have spares available;
(c) instruct men on how to use the plant on the method proposed for placing;
(d) have good access to all pours, particularly when using vibrating equipment on scaffolding;
(e) ensure that the concrete mix is suitable, that it can be compacted and completely fill the forms;
(f) have room for temporary chutes to be positioned and also have these light enough to be moved along to suit filling methods;
(g) take care to use the vibrators for their main function – to compact the concrete;
(h) make sure all plant is cleaned immediately on completion. Wet concrete is easy to remove from both plant and form-

work, but once it has hardened it becomes a costly task to clean off;

(i) formwork can be moved out of alignment while being filled with concrete, therefore time must always be allowed for plumbing and checking. It is important to have them correct after they have been filled;

(j) have someone checking the formwork for leaks or failures as the pour proceeds.

Suggested reading

1. Cement & Concrete Association. *Placing and compacting concrete.* (Man on the job leaflet).
2. Cement & Concrete Association. *Concrete practice.* 1979.
3. Barnbrook, G. *Concrete ground floor construction for the man on site. Part 1: for the site supervisor and manager.* (2nd edn) Cement & Concrete Association. 1975.
4. Barnbrook, G. Ibid. *Part 2: for the floorlayer.* Cement & Concrete Association. 1976.

8

Workability

Workability is an essential characteristic of concrete and unless this is correct for the particular requirement, a satisfactory end product will not be achieved. Concrete with the right amount of workability will have enough plasticity to allow it to be transported, placed and compacted.

This is the fundamental requirement of all concrete.

Workability is a function of the relative amounts of water and cement, known as the water/cement ratio.

Other factors that can affect workability include:

weather;	shape of forms;
aggregates;	reinforcement;
water;	method of placing;
mixing time;	strength required.

The weather

Exposure in hot weather and a combination of long hauls will cause loss of moisture and lower the workability, as will high winds. In contrast, rain will increase the moisture content and the workability. Consequently, all concrete in transit and placing needs protection.

Aggregates

Aggregates are the most variable materials used in concrete. Smooth rounded aggregate will provide better workability than aggregates that have sharp edges or a rough surface, such as crushed stone. The latter need more vibration to prevent them locking together.

When using crushed stone it may be necessary to increase the cement content to allow extra water to be added in order to obtain the desired workability without reducing the strength.

Aggregate sizes also affect the workability. As the size reduces, the cement content will need to be increased together with the water content. The reason for this is the total surface area of the aggregate to be wetted is greater with the smaller aggregate size. The fine and coarse aggregates should be proportioned to obtain the required workability, using a minimum amount of water.

Should aggregates be badly proportioned then excessive water will be needed and this will result in low strength and poor durability.

These matters are very much the concern of the engineer/mix designer but an awareness by the user will help to explain the performance of different materials and their effect on placing.

Water

The quality of water does not vary very much and most water can be used for making concrete; even sea water can be used for mass concrete, but not for any concrete that contains reinforcement. A good rule of thumb is that if the water is drinkable then it can be used for concrete.

Variation of quantity should be avoided and although this would seem to be a simple requirement, in practice it is rather difficult. The amount of water used should be the minimum necessary to give sufficient workability for full compaction of the concrete.

When deciding how much water is required some allowance has to be made for the amount of water that is probably contained in the pores of the large aggregate, in the sand, or on the surface of wet aggregates.

In practice, the mixer driver or ganger will soon become proficient in judging the workability of each mix and, therefore, be able to make the adjustments that frequently have to be made to suit the changing conditions.

For example if a water/cement ratio is 0.45 by weight, then the quantity of water for each 50 kg of cement (110 lb) should be 0.45 × 50 = 22.5 kg (5 gallons). This figure of 22.5 has to be reduced by the amount of free water contained in the aggregate.

The measurement of this free water is described in Chapter 12 on testing but as a general guide the average moisture content of sand is 5 per cent by weight and 2 per cent by weight of the coarse aggregate.

Mixing time

Should a mixing time be stated either in a specification or one appropriate to a particular type of mixer, the time given starts when all the constituent materials are in the mixer.

The biggest danger is with under-mixing, as this will affect workability; over-mixing can result in the excessive heating up of the concrete, so if delays occur then slow the mixer down. The aim of mixing should be consistency in timing; the colour of concrete can also vary if mixing time is inconsistent.

Shape of forms

The shape, height or size of the forms must be considered before determining the workability needed.

Narrow walls or balustrades, boot lintels and web-shaped beams are examples of items needing special consideration, for it is essential to have enough workability in the concrete to overcome the difficulties of placing and filling the forms.

Reinforcement

The main purpose of reinforcement is to provide the concrete's tensile strength, but this will be lost or reduced if it cannot be covered completely with well-compacted concrete.

When bending and assembling the link shapes, laps and bends of the reinforcement cannot be effected perfectly, so invariably it occupies more space in the forms than anticipated by the detailer. It is not uncommon to find that aggregate size and poker size have to be reduced in order to be able to place the concrete. It may even be necessary to take out (temporarily) some of the reinforcement to enable the concrete to be placed in the lower portions of the formwork, this steel being replaced as filling proceeds. Alternatively, the positioning of the steel can be altered to permit placing, but can only be carried out with the agreement of the design engineer.

Method of placing

Pumped concrete needs to have enough cement paste to lubricate the pipe line. By contrast, a conveyor needs the minimum workability to produce a cohesive mix in order to stay on the conveyor.

Required strength

Water assists workability but also reduces the strength of the concrete. All site personnel dealing with concrete placing must be made aware of the danger to concrete strength that results from indiscriminate adding of water after the concrete has been mixed.

Methods of testing workability

These are described in more detail in Chapter 12, with only the more common methods being covered briefly here.

It is important to regularly check fresh concrete in order to assess the practicability of being able to compact the mix and also to check if consistency is being maintained. Workability tests may also be used as an indirect check on the water/cement ratio that had previously been established in the laboratory. Should the correct proportions continue to be used at a constant workability, this will indicate that the water/cement ratio is also controlled.

The slump test

The slump test is simple to apply and is suitable for most concretes. Since it is a task that delays production, it is sometimes delegated to junior or untrained staff. Many problems arise from this test being carried out casually without proper instructions being given to the operative.

It must be appreciated that the test is one for consistency and it is preferable that it is performed by the same person using the same tools and following standard procedures.

Changes in slump may indicate changes in either materials, water content or proportions of the mixed materials. It provides, therefore, a very useful early warning system in the control of quality.

The 'slump' is the distance that a coneful of concrete slumps down when the cone is lifted from around it. It can vary from nil on dry mixes to complete collapse on very wet ones. When comparing workability batches of concrete should be around the same or within about 25 mm (1 in.) of the specified or acceptable standard.

There are three kinds of slump:

1. A true slump – where the concrete just subsides, keeping its approximate shape.

2. A shear slump – where the top half of the cone shears off and slips sideways down an inclined plane.
3. A collapse slump – where the concrete collapses completely.

It is possible for a true and shear slump to occur with the same mix but they should not be compared.

Only the 'true' slump can be measured. Therefore, if it shears first time a second test should be made. If this also shears it is probably due to the design of the mix and should be recorded on the test report. Collapsed slumps should also be recorded.

A slump test should be taken soon after the day's work has started. This will set a standard and give the mixer driver or ganger a visual check on that particular requirement. Further tests should be taken during the day, particularly if the mix is not looking, or working, right.

Each slump test should be made at the same time after mixing has been completed.

Any substitution or improvisation in the equipment used will defeat the objective of the test, which is to provide a comparative method of checking consistency.

Sampling

A proper sample of the mix must always be taken and must be representative of the batch to be tested.

Whenever possible the sample should be taken as the concrete is being discharged from the mixer or ready-mix truck.

As a test of workability only, a sample may be taken from the initial discharge, rather than in accordance with BS 1881 allowing slump tests to be taken after only 0.3 m^3 of concrete has been discharged. This will generally be more acceptable to the user rather than having most of the concrete discharged before a complete sample is permitted by BS 1881 (see Fig. 8.1, 'sampling and testing for slump').

Other tests for workability

Other tests, less suitable for site use, include the:

(a) compacting factor test;
(b) Vebe consistometer test;
(c) air content test.

These are described in more detail in Chapter 12.

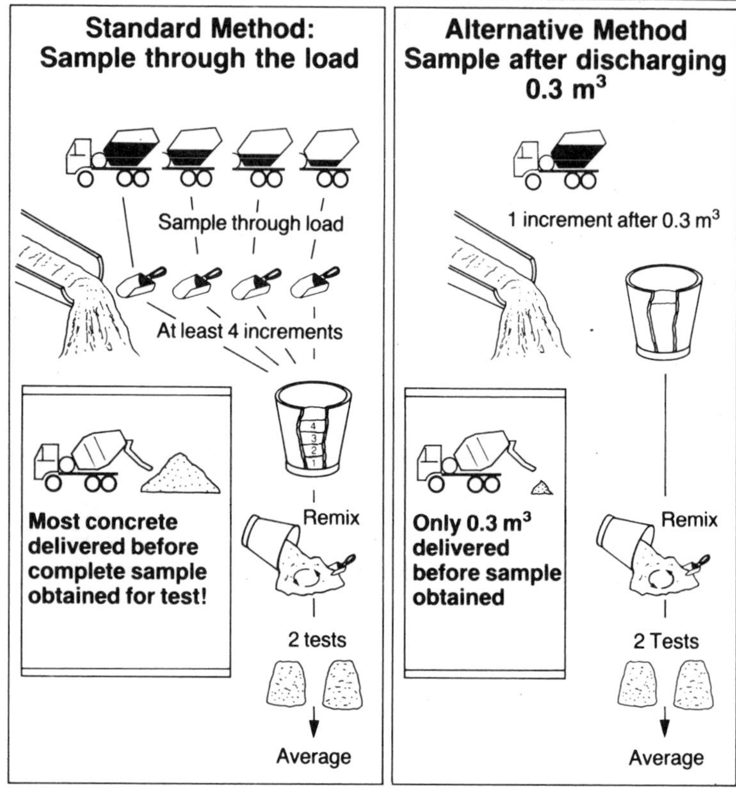

Fig. 8.1 Testing for slump – alternative methods of sampling

Summary

An experienced trained operative is able to recognise suitable workability and to spot immediately any variations occurring during the day. But it is also necessary to carry out routine testings and to have them recorded. These site tests support the visual observations. They do not measure any basic properties of the concrete but give a rough guide to the water content of the mix.

It is impracticable to test every batch so it is, therefore, important to visually check workability and also equally important to carry out the tests that are taken in a consistent and workmanlike manner to make sure that the following are taken into consideration:

(a) there is enough time to do the testing properly;
(b) clean and store the tools immediately after using;
(c) use the same processes with the same person whenever possible;
(d) carefully record the results;
(e) be able to interpret the results and take the correct action in the event of the tests not conforming to the specification requirements.

Workability is a fundamental property of concrete – unless the concrete can be placed and compacted to completely fill the mould or form and surround all the steel, ducts and other fixings, then it does not fulfil the basic requirement. Therefore, immediate action must be taken by the user to cease placing unsuitable concrete and to seek advice on the cause and remedy.

Suggested reading

1. Cement & Concrete Association. *Concrete practice.* 1979.
2. Cement & Concrete Association. *Construction guide: quality control of site mixed concrete*; (3rd edn) 1978.
3. Tattersall, G. H. *The workability of concrete.* Viewpoint Publications. 1976.

9

Joints in concrete

It is seldom practical or economical to place concrete continuously from beginning to end of construction, for the following reasons:

(a) economically the formwork has to be planned and designed for re-use;
(b) there is a limit to the amount of concrete it is possible either to produce or place in any one day;
(c) concrete has to be trowelled, tamped or finished;
(d) protection from the weather may have to be provided;
(e) the workforce have to go home.

All these create the need for some form of jointing.

Construction joints or, as they are sometimes referred to, 'day' joints, are an essential part of the construction process. These joints can be a source of weakness if they are not formed efficiently and located in the appropriate places.

The great majority of concrete construction projects are designed prior to the contractor's appointment and in these circumstances the designer is not aware of the construction methods the contractor will use. Consequently, it will not be known where and how the contractor will stop daily concreting.

This results in the contractor not having the opportunity to price for these 'day joints'. It follows that the contractor and his site staff need to give full attention and consideration to both the method and location of day joints, otherwise they can prove to be very costly.

There are three main types of joint employed to suit different situations:

1. Construction joints.
2. Water-tight joints.
3. Movement or control joints.

Construction joints

At construction joints fresh concrete is placed against concrete that has already hardened to provide structural continuity across the joint.

They can be formed of hardboard or timber, carefully fitted so that they can be removed as soon as the concrete has hardened sufficiently to remain in place. If the stop end has been cut around the projecting steel, as in a continuous slab or wall, then delay will make striking more difficult, particularly if the stop end material has any thickness, such as 20 mm plywood. Suitably supported hardboard or similar material, in short lengths, is preferable to timber, as it is easier to remove.

Permanent stop ends

Expanded metal may sometimes be used and left cast in, providing care is taken not to carry this to any part of the structure where it may eventually corrode (See Fig. 9.1).

Location

Stop ends should, wherever practicable, be either vertical or horizontal; on sloping pours they need to be at right angles to the slope.

Joints will always show in the finished concrete, so where appearance is important these joints should be arranged to fit in with any architectural features. Battens nailed to the formwork will ensure a horizontal line and disguise the joint, or a dovetailed feature can be formed between the two pours of concrete (See Fig. 9.2).

Forms need to be tight at construction joints because grout leaks are unsightly and difficult to remove; the loss of grout also disfigures and weakens the concrete. A thin strip of foam plastic on the lower form squeezed against the hardened concrete will aid the prevention of grout loss (See Fig. 9.2).

Forming a good joint

Fresh concrete can be bonded to hardened concrete without loss of strength if care is taken to do this properly. To obtain a good bond the hardened concrete must be clean and the fine material

LEEDS COLLEGE OF BUILDING

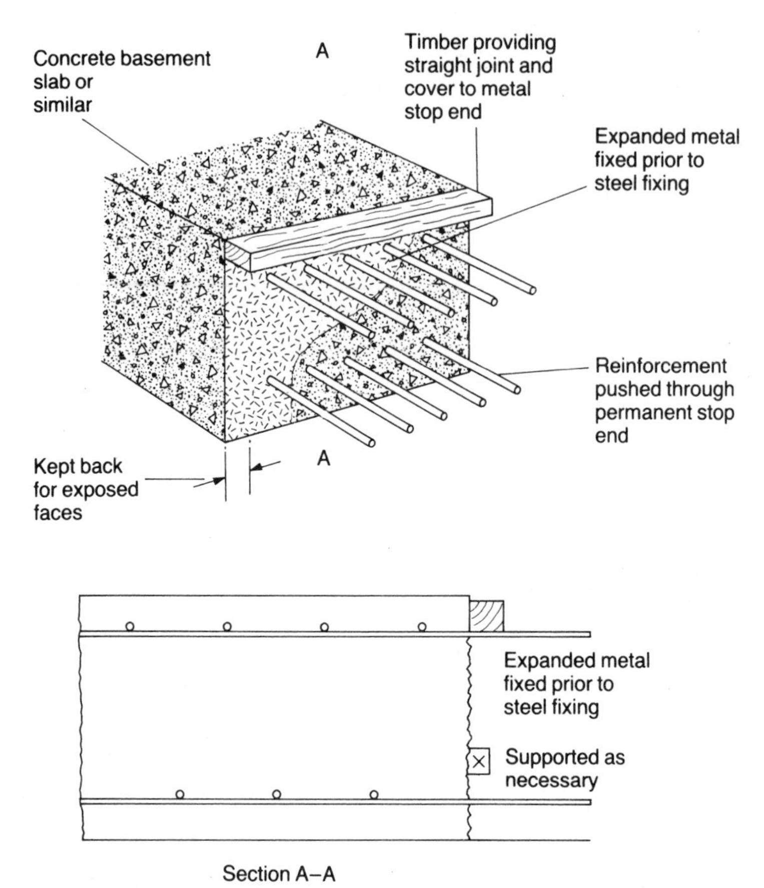

Fig. 9.1 Use of expanded metal as permanent formwork

removed at the face to expose the aggregate. This is because as the concrete is being vibrated, surplus water will rise to the surface bringing with it small particles of cement and fines. This can be reduced by using a drier mix for the top lift, but invariably some of the paste remains (called 'laitance'). Laitance having a high water content will probably bring with it any fine lightweight particles gathered from the aggregate. This is the weak porous material which has to be removed.

Removing laitance

There are several ways of removing the laitance, the objective being to expose the aggregate and to get a clean surface. By far

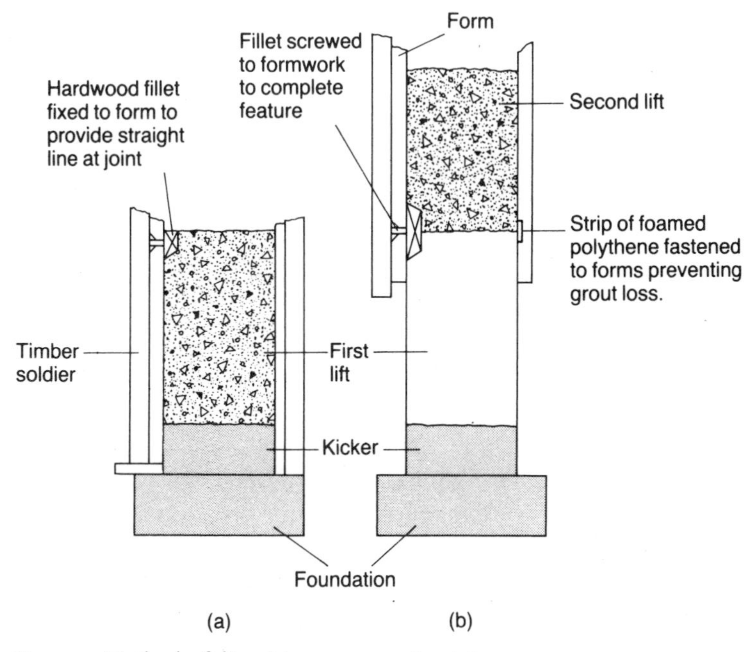

Fig. 9.2 Method of disguising construction joint

the best easiest and cheapest way is to brush off the laitance while the concrete is still fresh but has not hardened. This is a matter of judgement, as timing depends on the weather and also on the mix; rich mixes will stiffen more quickly than a lean mix. With practice the right timing can be selected; it is approximately 1–3 hours after the surface water has evaporated. It can be useful to arrange completion of a pour to occur giving time for this operation to be carried out successfully and conveniently.

Brushing must be gentle, care being taken not to leave projecting pieces of aggregate partly displaced.

Should compressed air be available, then this can be combined with water to remove the laitance very effectively (see Fig. 9.3). The concrete will have to be left for about 6–8 hours. Timing again is critical and care is needed not to disturb any large aggregate by using too much air pressure.

Hardened laitance, if it is still green, can be removed with washing and a wire brush, with a second wash after it has dried to remove any dust. When washing, protection will be needed to avoid laitance running down or over the previously cast concrete. If the

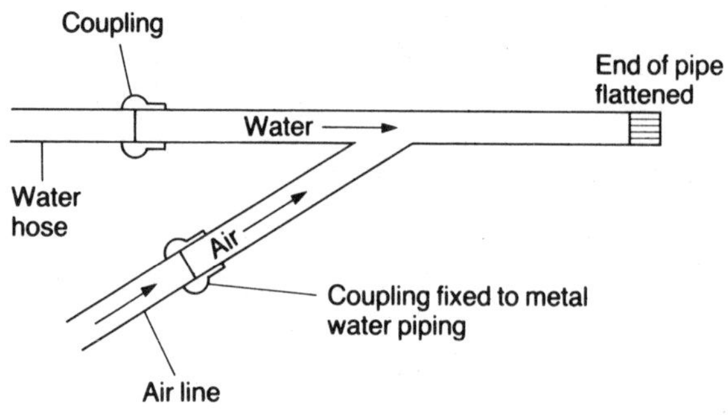

Fig. 9.3 Tool to remove laitance from hardening concrete

surface has hardened too much to be able to use brushes and water then mechanical scabbling will have to be employed.

The tools used for scabbling, like those used for exposed aggregate finishes, i.e. bush hammer head or a needle gun, can damage the concrete by shattering or loosening the coarse aggregate. Therefore, care needs to be taken and the concrete should be at least three days old. Wet or dry grit-blasting can be used for large surfaces.

Retarders or acids will also remove laitance. They must be used with great care to ensure none of the acid or retarder remains, otherwise a poor bond will result.

Having successfully produced a clean exposed aggregate surface it must be kept clean and the new concrete should be placed against this surface as soon as possible. Not all construction joints need this surface treatment. In horizontal joints that are always going to be in compression the flat finished surface is acceptable and quite satisfactory. If a roughened surface is required this can be provided by using a fine steel mesh or expanded metal loosely fastened to the forms, which is pulled off before the concrete has hardened.

These vertical stop ends are preferably located where the reinforcement will give the least trouble when removing the stop end and should be well made to avoid grout loss.

Movement and water-tight joints

These joints meet structural, thermal or water-tightness requirements. They are normally shown on drawings for location and described in the specifications as either joints between lifts or bays designed to permit movement but resist water; or joints to permit movement caused by heat, shrinkage or settlement.

Joints for water-tightness

To form joints between lifts or bays designed to permit movement but to resist water either entering, as in a basement wall, or leaking out of a water tank or reservoir, a 'water bar' is normally employed. They are usually made either from plastic or rubber in strip form with the cross-section shaped to be embedded in both sides of the joint either in the centre, the inside or outside of the joint (see Fig. 9.4). Where no movement is anticipated then the water bar can be in a rigid form, such as mild steel or galvanised steel, etc. Water bars are available in many forms to suit most situations and the reader is advised to obtain the advice and the very good technical and practical details that are offered from the leading manufacturers.

It cannot be over-stressed that the effectiveness of water bars very much depends on workmanship. They must be fixed rigidly in place before concrete placing commences and fixed to stay in that position when placing has been completed. Nails driven through the bar are not to be recommended for this purpose! Attention must be given to the complete compaction of the concrete that surrounds the water bar. This is particularly important as the presence of these bars tends to add to concrete placing problems because the space is often congested. If the concrete is poorly compacted, water can pass around the bar, thus defeating its object.

Joints to permit movement caused by thermal changes

Large elevations of buildings exposed to sun, bridges and roads are all examples of structures which are subject to expansion and contraction.

Movement can also occur where foundations are subject to different loadings, for example a tower block having a low level podium.

These joints are called movement or control joints and are formed by separating the structure, either by twin columns or plain wall faces, with a compressable material between.

When these joints are used externally the flexible material stops short of the exposed form, or the material is raked out and the

(a)

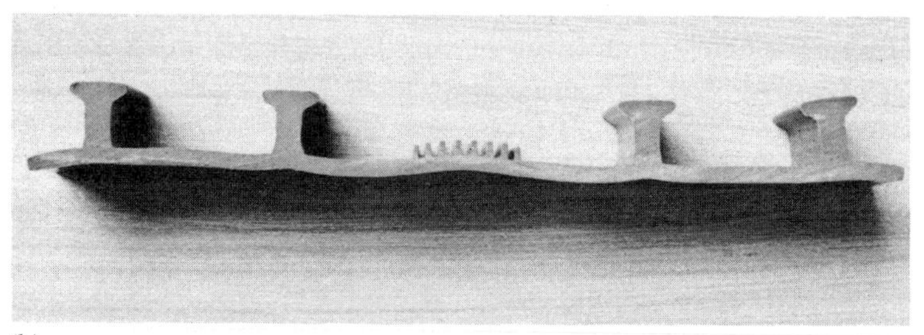

(b)

(c)

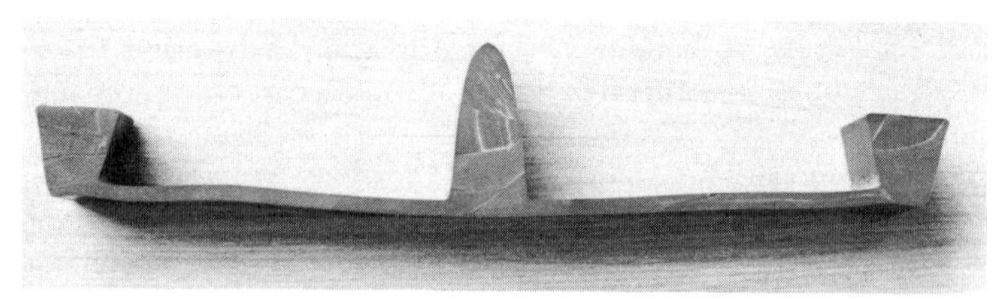

(d)

Fig. 9.4 Various types of waterbar viewed in section (a) internal dumb-bell type: (b) surface-type; (c) surface type with crack-inducing attachment; (d) surface type with integral crack inducer

joint is filled with a mastic, capable of providing water-tightness in addition to being durable and flexible.

Kickers for columns and walls

Kicker is a term commonly used to describe that concrete forming an upstand above floor level to position wall or column formwork for the next lift (see Fig. 9.5). Concreting kickers needs to be carried out with great care because the accuracy of the wall or column depends on the formwork being in the correct position after filling is completed. This can be made more difficult when the temporary fixing of the forms is not robust enough; some method of checking immediately after filling should be arranged.

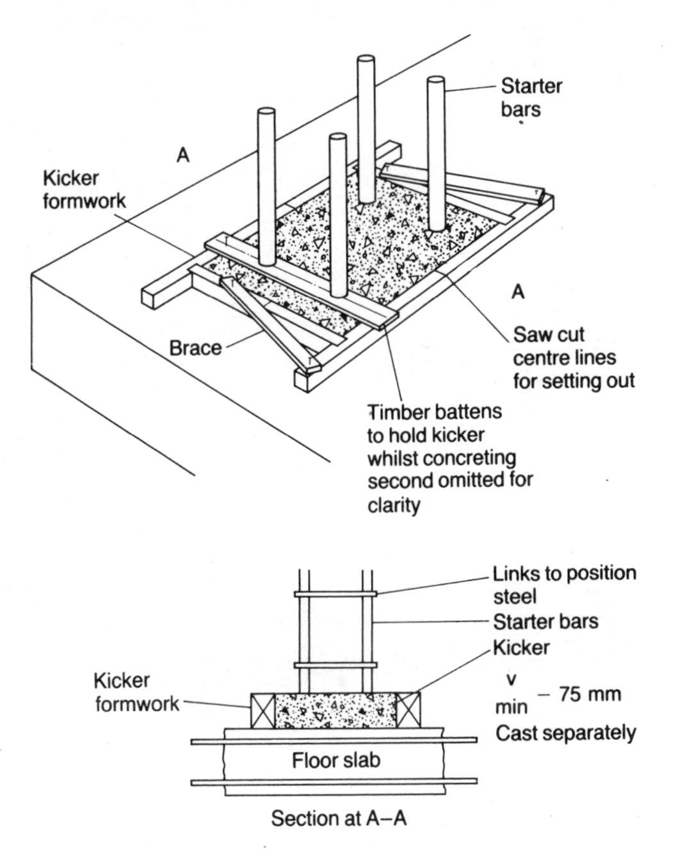

Fig. 9.5 Kicker for column

The importance of the strength of kicker concrete has now been recognised by designers and engineers, who recommend a minimum thickness of 75 mm.

This enables a fair sample of the concrete to be positioned in kickers which should also be vibrated; the formwork needs to withstand this compaction of the concrete.

Fig. 9.6 Kickers for columns showing starter bars for steel and kicker height of 75 mm (A)

Integral kickers

It is normally both acceptable and more economical to form kickers for columns and walls as a separate operation after the slab or wall has hardened, the steel is usually continuous from below and all that is necessary is to ensure that the jointing surface is clean with a wash from a hose pipe.

Kickers in water-tight constructions need different treatment. They are generally cast integral with the base slab or wall (see Fig. 9.7) and sometimes incorporate a mechanical water bar.

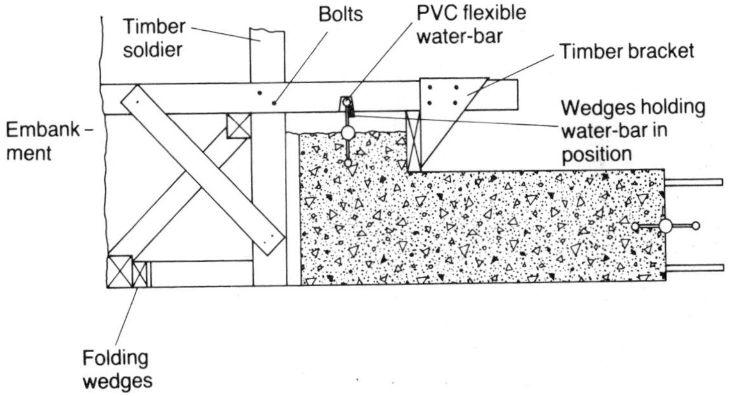

Fig. 9.7 Integral kicker for watertight basement construction

By casting these kickers together with the base concrete only one joint will be needed for the next lift. This should aid the water-tightness of the structure.

Good timing, formwork and supervision of water/cement ratio are necessary if the pressure of the kicker concrete is to be prevented from oozing out from the suspended formwork.

Special joints

In large civil engineering projects, roads, bridges and dams, etc., particular attention must be given to take care of the movement that occurs.

These joints will be detailed and designed for their function. The main concern will be to follow carefully the specification and to ensure good workmanship at these vital points.

Placing fresh concrete at joints

Regardless of whether a good bond is required, the placing of concrete on or against a joint needs special care and attention. Should the concrete be poorly compacted at the bottom of a lift in a wall or column, the result will be a weak and probably unsightly joint. Therefore, any dust, dirt, sawdust, nails or bits of tying wire must be removed. This can be done with a compressed air hose but if this is not available then a thorough brushing or washing out can be effective.

For washing out, a pocket needs to be provided in the lowest part of the formwork and a check carried out to ensure that it is re-fixed prior to concreting. Industrial vacuum cleaners can also be employed effectively for this purpose.

When placing the first layer of concrete some of the cement paste will be lost, not only when transporting but also when placing the concrete in wall or column formwork the paste will stay on the forms and the steel on its way down. To ensure a good joint the first batch of concrete should be richer than subsequent batches. This can be achieved by reducing the amount of large aggregate by 40–50 per cent in the first mix.

The site may need a small mixer for this purpose if concrete is generally being supplied as ready-mix.

The first discharge from a ready-mix supplier may also be a bit 'boney' (lacking in cement paste), as some paste will stay lining the drum. A small mixer will help this situation or alternatively the first barrowful out of the mixer may be put aside to be re-mixed and added later.

The first mix must be spread evenly and well compacted; small columns should be filled slowly, possibly by shovel from a skip or similar container.

Summary

1. Treat all joints with equal, if not more, care than the remainder of the structure. A chain is only as strong as its weakest link; likewise good concrete may fail to suit its purpose if jointed badly.
2. Aim to strike stop ends as soon as possible. They are much easier, and therefore cheaper, to remove while the concrete is still green.
3. When fixing a stop end remember it will be much more

difficult to remove once the concrete has hardened. Ensure the steel is firmly held and fill any gaps that provided tolerances when fixing.

4. Cleanliness is important when forming joints for watertight structures.
5. Good supervision and preparation are an essential part of successful jointing.

Suggested reading

1. Cement & Concrete Association. *Construction joints* (Man on the job leaflet).
2. Cement & Concrete Association. *Concrete practice.* 1979.
3. Monks, W. A. *Visual concrete – design and production.* Cement and Concrete Association, 1980.

10

Curing of concrete

The term curing is somewhat misleading as it implies there is something wrong with the concrete, thereby needing a remedy. The curing of concrete, like curing bacon, is simply to control the temperature and water content for a definite period, so that the concrete will be stronger and more resistant to stress and the elements; in other words, to preserve it.

When cement is mixed with water a chemical reaction commences called hydration which causes the hardening process and the concrete to develop its strength.

This strength development can only continue if the concrete is kept moist and at a favourable temperature, especially during the first few days after placing the concrete. Active hydration occurs during these first few days, therefore it is important for the water to be retained during this period. This means that evaporation should be prevented, or slowed down.

Evaporation will be higher in low humidity, in high winds, or if the temperature of the concrete is very high. The final concrete will be much improved if kept at a reasonably high and uniform temperature, avoiding excessive changes of temperature of both concrete and the surrounding area.

Cold and frost are not good for newly formed concrete neither are excessive heat or sudden changes in temperature. For example, cold water should not be sprayed, for curing purposes, on a hot day, over newly formed surfaces as the result is likely to be hair line cracking of the surface. The water needs to be at or near the temperature of the concrete. Most concrete will harden satisfactorily over a fairly wide range of temperatures. There is no need for further curing even in warm weather or if formwork on columns and walls is left in position for four days. However, particular care will be needed in frosty weather when expansion of the freezing water may cause fracture or spalling of the surface.

In hot weather the rapid escape of moisture at the plastic stage of the concrete can cause undue shrinkage, resulting in cracking.

Methods of curing

When making provision for the two extremes, frost and heat, the first factors to consider are those which influence the gain of strength, i.e. cement and water. Water is needed to provide workability and to commence the hardening process, and the proportion used, known as the water/cement ratio (w/c), must be sufficient for these two tasks. The ratio will need to be varied to suit climatic variations.

In cold weather the water/cement ratio needs to be kept to the minimum necessary to provide the desired workability. Since cement and water react more rapidly as the temperature increases, it is an advantage to use warm water to assist early setting as a precaution against frost damage. In hot weather the water will have a tendency to evaporate more rapidly because the speeded up hardening process produces more heat. To increase the water/cement ratio without taking additional precautions would only create problems of cracking and loss of strength.

Mix adjustments to assist curing

When concreting in hot weather the shading of the aggregate stockpile and of the concrete where it is being placed are sensible precautions. Cooled water can be used both for mixing the concrete and also sprayed on the aggregate to reduce temperature.

In cold weather the water and aggregate can be heated, but cement should not be heated. To avoid problems the aggregates are placed in the mixer first, followed by the hot water and finally the cement. Alternatively, steam can be used.

Accelerating admixtures are marketed as an aid to cold weather concreting. Where these are named and specified, the user will be well advised to read the maker's instructions and carefully supervise their use. Unfortunately, site personnel abuse these materials by thinking that if 2 per cent is good then 4 per cent plus will be marvellous! Defective concrete can result from excessive dosage or uneven distribution of admixtures throughout the mix.

The advice given in the workmanship clauses of the Code of Practice CP110 is very sound when it recommends that additives should not be used as a replacement for good concrete practice and should never be used indiscrimately. There is already a considerable number of variants to take into account, for example, the weather, the temperature, the cleanliness of the materials, the amount of water, the mixing time, the quality of the materials, etc., so the use of additives will only complicate the task further. For the production of ordinary structural concrete their use should not be necessary.

Admixtures may be required in 'special structural concrete' but should only be used with prior approval of the engineer. In these circumstances, both the engineer and the user should be sure that the admixture contains no chlorides; if so, what amount and whether the admixture leads to the entrainment of air when the recommended dosage is used.

Should two or more be specified for one mix, the manufacturer of each should be consulted to be sure they will not have a harmful effect on mixing.

Particular care is needed if admixtures are used together with special cements.

Controlled use of retarders in hot weather and air entrainers in concrete that has to withstand cold conditions are examples of the correct application of admixtures.

Water spraying

Water spraying is a method of curing and must be carried out thoroughly. A continuous fog spraying or sprinkling of water is required. To saturate intermittently and then allow the concrete to dry will do more harm than good.

The water needs to be at a temperature close to that of the concrete surface and of an adequate supply. Sea water can be used for curing providing construction joints are washed with fresh water before placing fresh concrete.

When curing it is important to ensure that the hose or other means of supplying water is independent and additional to the general supply. This will reduce the risk of the curing water being used for other site needs and then not being reconnected.

Wet coverings

Surfaces can be effectively cured by using hessian or other absorbent materials, providing they can be kept continually wet and are uniformly positioned.

Coverings can be laid over the slab as soon as it is hard enough to withstand any surface damage but they should be supported at the slab edges (to prevent damage) and lapped and fastened to prevent displacement in high winds.

On roof slabs or areas where discolouration is not important, sand spread evenly and kept damp is a useful curing method, but the cost of spreading and recovery must be taken into account.

Waterproof paper

Waterproof paper can also be used on unobstructed horizontal surfaces. If the widest practical width is used and sections fastened together with an approved waterproof adhesive, with a lap of 100 mm, these covers can be rolled on early hardened concrete and will be re-usable several times if handled carefully.

In addition to curing the concrete paper gives protection to the concrete from damage by other construction activities and also from the effect of the sun. It will need weighting down or fastening in some manner as a precaution against the wind. The paper can be repaired by patching and using suitable adhesives and finally used double when its condition has deteriorated.

Two layers of waterproof paper clipped together with straw in between is an effective way of protecting concrete in cold weather.

Polythene sheeting

This can be used in a similar way to waterproof paper, providing cost and supply are comparable.

White sheeting will reflect the sun's rays and will be useful in warm weather when black sheeting should be avoided. The black sheeting will absorb heat so has advantages when used in cold weather.

Polythene sheeting can cause discolouration or mottling of the hardened concrete because of condensation or uneven distribution of water on the underside of the covering. This will limit its use to base slabs that are going to receive a topping or screeding for floor finishes.

Polythene sheeting, like waterproof paper, should be laid as soon as possible without damaging the surface and it will provide some protection against frost if stretched on, or over, light timber frames. This will trap a layer of static air between the sheeting and slab, but for it to be effective the edges need to be well secured.

Fig. 10.1 *In-situ* supported ground-floor slab being cured with polythene sheeting

This will also prevent wind from blowing between the freshly placed concrete and the underside of the sheeting, otherwise the wind will dry the top surface too quickly and cause skin cracking.

Polythene sheeting can be obtained in various thicknesses for a wide variety of purposes, including covering light frames to make temporary portable shelters or covers. It can also be obtained reinforced with nylon netting if needed to be more durable.

Vertical surfaces of in situ columns can be protected with lightweight sheeting to retain the moisture, and if this is left on it will also keep the surfaces clean and clear of grout runs from subsequent concrete pours.

Pre-cast concrete having special exposed aggregate surfaces can be protected by shrink wrapping with polythene.

Spray-on membranes

Liquid membrane-forming materials containing waxes, resins, etc. are applied by spraying horizontal or vertical surfaces to retain moisture or retard evaporation.

The effectiveness of sprayed membranes depends on the quality of the material and the thoroughness and timing of the application.

The success or failure of a membrane-curing application will not become apparent until long after the concrete has hardened and been exposed to extreme weather.

It is difficult to assess the quality of the material before it is applied so it is essential to obtain it from a reliable supplier or specified manufacturer. Some curing membranes can be completely useless.

The material should be sprayed evenly as soon as the free water has evaporated and the sheen disappeared from the surface. If the application is delayed, the curing material will be absorbed by the surface pores of the concrete, thereby reducing its effectiveness and staining or discolouration may be caused.

Normally only one coat is needed, but should a second coat be considered advisable this will be better applied by passing the spray nozzle in sweeps at right angles to the first application.

On formed surfaces the material should be applied immediately the forms are removed. Sprayed membranes containing a dye are helpful to enable the operator to ensure full coverage and to know where to restart after stopping for refills or cups of tea.

This colouring, if white or aluminised, has the additional value of reducing the heat gain by reflection of the sun rays. These compounds are applied by hand or power-driven spray equipment, which, if properly maintained, provide a smooth, even and continuous texture. Complete coverage of the surface and sides must be attained as even very small gaps or holes will permit evaporation. Manufacturers' instructions are important and should always be followed.

Curing compounds may prevent a bond between the hardened concrete and any screed or wet concrete applied later. Therefore, they should not be used in these circumstances. The majority of sprayed curing agents disintegrate with time and exposure to ultraviolet light, but their effect on bond reduction will last longer if used internally on floors or factory slabs, so other means of curing should be employed for such situations.

Use of heat for curing

It has been stated that heat accelerates the hardening process of concrete. In cold weather curing time can be reduced in a number of different ways, including space heating, electrical curing and steam curing.

Space heating

Even on large or scattered sites, the area ready for concrete can be temporarily enclosed by the use of screens, tarpaulins or plastic sheeting. This enables oil or electric space heaters or fires to provide and maintain in this limited area a temperature which will keep the concrete at least above 5 °C. Any further improvement in temperature will reduce the curing period and possibly permit the earlier release of formwork. Where work has progressed sufficiently the building itself and the scaffolding can provide the means to enclose space for heating purposes.

Electrical curing

Electrical curing is replacing braziers as a form of accelerated curing, primarily in pre-cast factories, to assist in the reduction in the number of moulds. Low voltage electricity is used and by embedding wire heater elements between plywood laminations, thermal efficiency is high and the method clean and easy to use. The electricity tariff is cheaper at night and this is when the heat is most useful. Repeat items on sites or in situ slabs can be cured in this manner.

Some form of thermostatic control is necessary, otherwise normally straight members may finish banana shaped.

Steam curing

Steam curing is a favourable method of curing. Pre-cast bridge beams, or similar structural members are covered with tarpaulins and pipes fed underneath conveying moist hot steam. This provides ideal conditions for maintaining moisture in the concrete and thereby accelerating curing.

Curing periods

The amount of time needed for curing will vary depending on the weather and the mix of concrete and also to some extent the sectional size of the member.

As a guide Table 10.1 shows variations under changing weather conditions together with changes of types of cement.

Table 10.1 Curing damp where average temperature of concrete exceeds 10 °C

Weather	*Low heat Portland cement*	*Ordinary Portland cement*	*Rapid hardening cement*
Normal	4 days	2	2
Hot or with drying winds	7	4	2
Damp		No special requirements	

Curing roads and paved areas

Large areas of exposed concrete are particularly vulnerable to the effect of changing weather conditions and no slab should be left unprotected for more than a short time after finishing. If it has not been sprayed or alternative curing procedures arranged, it should at least be shielded from the sun and wind by low tents of fabric just clear of the surface and closed around the edges to prevent excessive wind damage.

Curing floor toppings

Separate floor toppings such as granolithic or sand and cement screeds can give problems if allowed to shrink before being strong enough to stay bonded to the structural concrete. If allowed to dry out too quickly they will shrink mostly at the surface and cause curling and lifting.

Waterproof paper or polythene sheeting is preferable to hessian because damp hessian may stain the finished surface.

Floors should be cured for at least a week or longer in cold weather. They should be protected when laid in buildings under construction by temporarily sealing as many doors and windows as possible to eliminate any draughts.

To allow curing to be effective, turning on the central heating should be gradual and delayed as long as possible; certainly several weeks. The longer the better.

Curing of floor toppings must be organised and supervised properly so that it is continuous. The floor is often used too soon by construction traffic and the curing is halted or neglected.

Curing in cold weather

Curing will take longer in cold weather and the concrete will take more time to gain strength. Under such conditions the formwork becomes important. On suspended slabs it will need to remain in position longer and have good insulation; for example, timber or an expanded polystrene lining to metal forms.

Wall and column formwork will also need to remain in position for a longer period. Should concrete have to be placed when the air temperature is below 2 °C special care will be needed to make sure the temperature of the concrete does not fall below 5 °C and that it will not go below this temperature while reaching its full compressive strength.

There is a considerable amount of misunderstanding on the subject of concreting in cold weather, some of which is caused by specifiers forbidding the placing of concrete in freezing or near freezing conditions. The decision on whether or not to place concrete should be based on the temperature that can be obtained and maintained in the concrete.

Concreting is carried on in Iceland and Canada at very low air temperatures. In Iceland there is a resource that is denied to other countries: a hole can be drilled on or near the site that will provide a source of hot water. This water together with heavily insulated forms allows construction to continue under extreme weather conditions. Sites need not call a halt when temperatures fall providing the following provisions are made to meet winter conditions:

- (a) the water can be heated;
- (b) formwork is of timber or insulated metal forms;
- (c) aggregates can be covered and also heated by steam coils or similar;
- (d) the placement area can be shielded from cold winds by temporary screens;
- (e) butane flame can be used to clear off any ice from the fixed reinforcement or formwork;
- (f) formwork can be covered overnight with insulating mats or electrically heated tarpaulins. These covers can be used later to cover the placed concrete;
- (g) rapid hardening cement is available if required;
- (h) space heaters or braziers are used below suspended slabs to combat the effect of frost. Corners and edges need extra attention.

The main objective of these precautions is to conserve the heat that is given off during hydration. During the period of freezing, or near freezing, the loss of moisture by evaporation will be very much reduced, so water curing will not be necessary nor advisable.

It is the temperature of the concrete that is critical not the actual temperature.

Curing in hot weather

Providing the hardening and curing of concrete is understood concreting in hot weather will not present any great problems.

Fresh concrete needs to be protected from the drying effects of sun and wind. Should the surface area of the concrete stiffen more rapidly than the rest of the concrete, tension will occur and will break the surface. This is known as plastic cracking.

The cracking may be random but on occasions it occurs over obstructions as when large aggregates bridge over reinforcement and water evaporates rapidly causing settlement to be uneven. The resultant cracks follow the line of reinforcement.

This cracking can be closed by re-tamping or re-vibrating if carried out promptly. It can be prevented or minimised by;

(a) dampening the sub-grade;
(b) wetting the aggregates if dry and absorbent;
(c) providing windbreaks to reduce the effect of drying winds;
(d) lowering the concrete temperature by cooling the aggregates and the mixing water;
(e) protecting the concrete with wet coverings should any delay occur between placing and finishing;
(f) using fog spraying to minimise evaporation for the important first few hours after finishing, cover as soon as practical.

The main objective is to prevent or reduce rapid evaporation of the water in the concrete.

Effect of curing on surface finishes

Curing will affect the texture, colour and durability of concrete. Where appearance is important special care is required to be sure that surface finishes of the same type are based on the use of identical methods and materials. Sprayed membranes can stain a special finish and changing or even repairing formwork will often alter the texture and colour.

Any concrete allowed to dry quickly will be lighter than concrete kept moist by leaving the forms longer in place. However, this colour difference will become less noticeable with time.

Special finishes

Special finishes will deteriorate in exposed conditions if attention is not given to compaction and curing of the concrete which provides cover to the reinforcement. These surfaces should be protected against excessive evaporation, which results in a weak porous layer susceptible to attack by frost, or moisture penetration thus damaging the reinforcement. Special coloured or white cements can be cured by ordinary methods, providing care is taken to avoid staining of the surface. Waterproof paper or polythene sheeting is safer than hessian and should be fixed to frames holding it clear of the concrete to reduce the possibility of drips or runs.

Summary

The durability and appearance of concrete depend very much on the selection and application of the appropriate curing method.

Operatives need to be instructed as to the purpose of curing and why it has to be continuously and conscientiously performed. Before concreting is commenced the curing methods have to be decided and sufficient men, materials and time allocated to provide adequate curing – last minute improvisations or intermittent soakings and drying can do more harm than good.

Externally exposed concrete needs more attention than interior members – badly cured, they will deteriorate quickly.

Caution is needed to ensure that the correct curing compounds are used with the recommended methods of mixes.

Curing is a vital part of the concreting process and all site staff should be made aware of this.

Suggested reading

1. Cement & Concrete Association. *Curing concrete* (Man on the job leaflet).

2. Cement & Concrete Association. *Concrete in cold weather* (Man on the job leaflet).
3. Shirley, D. E. *Construction guide: concreting in hot weather.* (4th edn) Cement & Concrete Association. 1981.

11

Concrete admixtures

An admixture is a chemical addition to the essential components of concrete.

Admixtures are unlikely to make poor concrete any better and are not substitutes for good mixing. They can aid compaction, make placing easier, give faster striking times or alter the hardened concrete.

The correct use of an admixture is a way of producing a more durable concrete or assisting in obtaining better concrete under extreme or unusual conditions.

Typical uses of some admixtures are given in Table 11.1.

When admixtures are specified the following check list will help to ensure that they are used correctly.

1. Has the selected admixture been delivered correctly? Never use one from an unmarked container and always check that the job specification permits its use. Also check and keep to recommended storage conditions.
2. Ensure that the correct dosage is used for each batch. It can be controlled more reliably by using a dispenser, generally available on hire from the admixture supplier. Using a dispenser will reduce the possibility of overdosing. Before commencing mixing concrete each day, the discharge from the dispenser should be checked for accuracy and at the end of the day it must be thoroughly washed out.
3. It is important for the admixture to be dispersed evenly throughout the concrete and the most effective way to do this is to put the liquid admixture into the mixing water before it goes into the mixer. Where this is not possible, the concrete will need to have a longer mixing time to ensure even distribution.

4. Carefully watch for changes in the aggregate gradings and water content. If the mix has to be changed the admixtures may also have to be adjusted.

Table 11.1 Typical use of some admixtures. (Asterisks indicate where they are most used)

Use	Set retarder	Set accelerator	Strength accelerator	Water reducer	Air-entraining agent	Water repellent	Superplasticiser
High early strength		*	*	*			
Hot weather	*			*			
Cold weather			*				
Frost resistance				*	*		
Workability aid				*	*		*
Chemical resistance				*	*		
Early formwork removal			*				
Pumping				*			
Precast cladding		*	*		*	*	
Improved surface finishes					*	*	
High strength				*			
Mortars					*		
Permeability reduction				*	*	*	
Waterproofing				*	*	*	
Mass construction	*						

Types of admixture

Set retarders

Set retarders retard the setting time of the concrete. They have limited scope in the United Kingdom but may be helpful under the following conditions:

(a) in warm weather (20°–25 °C) to prevent early 'going off' and loss of workability;
(b) for large pours of concrete taking several hours, to maintain workability;
(c) when slipforms are being used they prevent cold joints which might arise because of the longer period needed between pours;
(d) where long delays between mixing and placing occur. For example, traffic delays or long journeys by the supplier. These can be more serious in hot weather, or with a concrete having a high cement content.

Trial mixes are necessary with set retarders, for although 7- and 28-day strengths are not likely to be affected, except by overdosage, strength at 24 hours to 48 hours may be reduced. This can influence formwork striking times.

Set accelerators

Set accelerators are used to speed up the rate of chemical reaction between cement and water, thereby causing the concrete to stiffen sooner and to develop early strength quicker.

This feature in concrete was once achieved with calcium chloride, but it has now been established that even small amounts of chlorides in concrete increase the risk of corrosion to the embedded reinforcement and consequently has been prohibited.

There are chloride-free accelerators available but they are often expensive and not so effective as calcium chloride. It is advisable, therefore, to follow the recommendations of the Code of Practice that admixtures should not be used as a substitute for good concrete practice.

Accelerators are not effective in mortars because the quantity of cement in the joint is too small to gain anything from the minor gain in heat acceleration.

No accelerator has anti-freeze characteristics so covering is always necessary to protect brickwork from frost.

Strength accelerators

Strength accelerators are used to obtain high early strength mainly by precasters desiring early formwork removal and re-use.

Water reducers

Water reducers are surface-acting chemicals having a lubricating effect that allows a reduction of water for a given workability; this may also allow a reduction of the cement content. Care must be taken in their use and due consideration given where a minimum cement content has been specified for durability reasons. Trial mixes and engineer's approval are advisable.

Air-entraining agents

Non-technical users of concrete, having been trained to exclude entrapped air from the concrete, find difficulty in appreciating the value of using an air-entraining agent. The point is that normal entrapped air is in the form of irregular bubbles creating voids in the hardened concrete. By using a controlled amount of air entrainer (usually producing about 5 per cent of air by volume) millions of tiny bubbles of uniform size are distributed homogeneously throughout the concrete mix. These minute air bubbles, acting like ball bearings, greatly improve the workability.

They permit a reduction in water content and compensate the mix designer for the loss of strength due to the added air. Workability from the entrained air improves all concrete and will also improve the cohesion of harsh mixes and reduce segregation and bleeding.

The main reason for adding this agent is to increase the hardened concrete's resistance to frost damage and de-icing salts.

As with most other admixtures, trial mixes are necessary to establish the correct amount needed combined with strict and consistent control to cater for any variations in materials or weather conditions.

Water repellents

The normal method of making concrete virtually impermeable to water is to design it to be crack resistant and to make the section of a dense thick concrete.

There are occasions when this is not possible; for example, when using a porous decorative aggregate or rendering a leaky basement. In these cases the use of a water-repellent admixture can assist in waterproofing.

Water repellents are sometimes added to reduce the need for frequent cleaning of buildings where they are made dirty by rain washing dirt into the surface.

These admixtures need careful supervision and should only be used when specified.

Superplasticisers

A fairly recent addition to the range of admixtures are superplasticisers which can increase a slump of 75 mm to a slump of 200 mm without affecting the strength or causing segregation or bleeding.

The mix has to be designed to accommodate the superplasticiser and its use can add £2–£3 to the cost of a cubic metre, but this can be viable in the following situations.

1. When added to a mix (this is best done at the site because the increased workability only lasts for about 60 minutes) the concrete can be described as 'flowing' concrete. This makes it very useful where reinforcement is particularly congested or sections of pre-cast units are difficult in shape. It reduces the need for vibration and speeds up the placing process. The 'flowing' property is valuable for large areas of concrete, such as slabs, that benefit from the ease of placing and increased speed of operation.
2. They can also produce high strength concrete by acting as a water reducer. As much as 30 per cent of water can be left out compared with about 10 per cent when using a normal plasticiser. High early strengths can be achieved permitting earlier formwork removal. The user must obtain further information before attempting to employ superplasticisers. There is also a special method for measuring the slump of these high slump concretes.

Summary

Admixtures are not a substitute for good concrete practice. Mixes must be properly designed, the concrete must be carefully conveyed and then placed and compacted with care.

They have to be selected carefully, specified in detail and measured accurately in dosage.

They need to be properly and clearly marked, stored under conditions recommended by the manufacturers and kept away from any materials that could be used in error.

Overdosing has to be avoided because it can be harmful to the concrete. Supervision and quality control has to be of a higher standard than for normal concrete practice.

Suggested reading

1. Pink, A. *Winter concreting.* (3rd edn) Cement & Concrete Association. 1978.
2. Cement & Concrete Association. *Concrete admixtures* (Man on the job leaflet).
3. Monks, W. *Visual concrete – design and production.* Cement & Concrete Association. 1980.
4. Roeder, A. R. Admixtures for concrete. *Site Management Information Service Paper No.* 73. CIOB. 1978.
5. Rixom, M. R. (Ed). *Concrete admixtures: use and applications.* Construction Press. 1977.

12

Quality control – methods of testing

Makers of concrete, whether on site or at a ready-mix plant are, essentially, manufacturers of a structural material. However, its production is subject to a number of variables, such as the quality and composition of the materials used and site conditions and weather. Operatives and supervision may also change during the period of a contract.

The specifier or client will need assurance that the concrete will meet strength and durability requirements and to this end a number of tests are available.

Achieving quality control

The amount of time, money and effort given to the control of quality will depend on the importance of the project. With a job of minor importance an experienced man just observing the concrete should be sufficient to ensure reasonable consistency. However, some formal system of control will be necessary on most jobs, and such a system may be classified as:

(a) moderate control;
(b) good control;
(c) close control.

Moderate control

A site seeking 'moderate control' would:

1. Store the materials correctly.
2. Have the fine aggregates tested for silt content before the work starts and whenever there seems to be a reason to

do so; the cement will be either batched by weight or, by using a fixed number of whole bags, for each batch of concrete.

3. Have suitable gauge boxes or other measuring containers where aggregates are batched by volume. Whenever possible it is preferable to batch cement and aggregates by weight.
4. Check the moisture content of aggregates before work starts and whenever a change is detected.
5. Check the quantity of water added at the mixer whenever in doubt.
6. Control workability (after an initial slump test) by visual inspection and take slump tests where necessary.
7. Make test cubes which are stored on site but sent for independent testing.
8. Supervise all operations adequately.

Good control

Where there is good control:

1. The sieve analysis and moisture contents of the coarse and fine aggregates are determined at commencement and at regular intervals as work proceeds.
2. The silt content of the fine aggregate is examined regularly.
3. Aggregates and cement are accurately batched by weight.
4. The amount of water added at the mixer is checked regularly.

Close control

This implies that all the requirements listed in the previous two sections are met, in addition to which:

1. All tests are made at least daily.
2. Cubes and other tests are made to accepted standards and either tested locally or on site.
3. Weigh batching plant and water measuring appliances are calibrated regularly.
4. Proportions of the fresh concrete and all other tests required are determined by approved methods.

5. The selected workability tests are carried out regularly and all work stages are closely and consistently supervised.

'Good' and 'close' supervision normally needs the addition of a technical man on site, together with accommodation for laboratory and testing equipment. This need only be part of an office or store, but on a large site it will require a properly fitted hut with heating for cold weather.

The site man will need to carry out, or supervise, the following basic tests:

(a) cleanliness tests on coarse and fine aggregates;
(b) slump tests (workability);
(c) cube making and storing (strength).

Silt test for sand

The silt content of sand should not exceed 8 per cent by volume. To carry out the test on site all that is needed is a 500 g (1 lb) jam jar, a teaspoonful of salt, water and a rule. The procedure is to have about 50 mm (2 in.) of sand in the jar and pour in salt water until the level is about 25 mm (1 in.) above the sand. Shake the jar well. Leave to stand for 3 hours, then measure the depth of silt that has settled on top of the sand; this should not be more than 3 mm (1/8 in.). To confirm that a load of sand is suspect, rub a sample in the hands. If the hands become soiled then the silt level is likely to be too high. The field settling test can, with experience, give a guide to excessive silt or clay in about 10 minutes. This may be enough to save having a doubtful load tipped, the short wait being preferable to having to remove an unsatisfactory load.

Test for workability

A simple means of evaluating workability is the slump test. Slump is the distance a coneful of concrete slumps down when the cone is lifted.

It has its limitations for very dry mixes; a compacting factor test should then be used. Slump tests enable a comparison to be made of workability of various batches of concrete. Providing materials and gradings are reasonably uniform, slumps should be within 25 mm (1 in.) of the intended value. Any larger variations will need investigation, particularly in regard to the water content of the aggregates. Checks should be made of the water gauge or any other reasons that have changed the water/cement ratio.

Slump tests need to be taken as often as is necessary to ensure that workability is consistent.

One test should always be made soon after starting the day's work. Any other tests made during the day should be taken at the same interval following mixing.

Equipment required for a slump test

1. A standard slump cone – 300 mm (12 in.) high with a bottom diameter of 200 mm (8 in.) and a top diameter of 100 mm (4 in.).
2. A steel tamping rod – 16 mm (5/8 in.) diameter, 600 mm (2 ft) long, with both ends rounded.
3. A waterproof baseplate, about 450 mm (8 in.) square – this can be made from 20 mm exterior quality plywood covered with 18 gauge steel plate.
4. A scoop.
5. A steel float.
6. A rule.
7. Cleaning rags.

All equipment must be cleaned after use and not left lying about on the site in the open.

Sampling

To enable accurate comparisons to be made samples are best taken by the same person and in the same manner throughout the day.

Preferably four roughly equal parts should be taken as the concrete is being discharged from the mixer or ready-mix truck and then mixed together to be representative of the whole batch. Alternatively, as mentioned earlier, all the sample may be taken when 0.3 m^3 has been discharged from the ready mix truck. The size of this sample will depend on whether cubes are to be made at the same time.

Should it be necessary to take a sample from a discharged heap, then extra care is necessary in order to obtain a representative sample. At least six portions should be taken, by digging deep into the heap so avoiding taking any from around the edge of the pile where large stones may have possibly accumulated.

Fig. 12.1 Equipment for the slump test

Test procedure

1. Make sure the cone is clean, free from hardened concrete and dry inside. Stand it on the base plate, which must also be clean.
2. Stand with feet on the foot rests.
3. Using the scoop, fill the cone to about one third of its height and rod this layer of concrete exactly 25 times using the tamping rod.
4. Add two further layers of equal height (each about 100 mm deep), rodding each one in turn exactly 25 times, allowing the rod to penetrate through into the layer below. After rodding the top layer make sure that there is a slight surcharge of concrete, i.e. some concrete sticks out of the top.
5. Strike off the surplus concrete using the steel float (see Fig. 12.2).

Fig. 12.2 Striking off and smoothing the top, using the steel float

6. Wipe the cone and base plate clean, keeping on the foot rests.
7. Take hold of the handles and pushing downwards remove feet from the foot rests.
8. Lift, very carefully, the cone straight up, turn it over and put it down on the base plate next to the mould of concrete. As soon as the cone is lifted the concrete will slump to some extent.

Fig. 12.3 With feet still on the foot-rests, wipe the cone and base plate clean

9. Rest the tamping rod across the top of the empty inverted cone so that it reaches over the slumped concrete.
10. Using the rule, measure from the underside of the rod to the highest point of the concrete (to the nearest 5 mm). If the distance is, say, 50 mm (2 in.) this represents a 50 mm slump.

Fig. 12.4 The cone must be lifted carefully and quite straight

Types of slump

Slumps are of three general types:

1. A true slump – where the concrete just subsides, approximately keeping its shape.

2. A shear slump – where the top half of the cone shears off and slips sideways down an inclined plane.
3. A collapse slump – where the concrete collapses completely.

Fig. 12.5 True, shear and collapsed slumps

Both true and shear slump can happen with the same mix, but one must not be compared with the other. The only one that is of any use is the true slump.

With a shear slump, a second test should be made in order to obtain a shape more nearly the 'true' slump. If this also shears, it is probably due to the design of the mix and this should be recorded. Similarly, collapsed slumps should be recorded as 'collapsed slumps'.

Records should be kept of slump tests giving the date; where the concrete was used; the amount of slump and description if not a 'true' slump.

The variation normally permitted is usually plus or minus 25 mm. Advice of the specifier or engineer should be sought when slumps exceed this tolerance.

Cube tests – tests for strength

It cannot be overstressed that the cube test is a test limited to checking the strength consistency, or otherwise, of fresh concrete. The strength and durability of the concrete in the structure will depend on how it is transported, placed and cured.

To produce good concrete, together with obtaining good cube

results, is only one part of the process. This can be ruined if the concrete is poorly transported, compacted or improperly cured.

Nevertheless, cube testing is a very important function and must not be left to an untrained amateur. The ready-mix industry estimates that £3,000,000 is wasted annually on meetings and expensive tests on hardened concrete due to tests that are carried out inefficiently on sites. Therefore, it is important that before appointing anyone to make cubes the operator should be provided with not only the correct equipment and facilities but also to know why the tests are taken and how to carry these out to the British Standard or method described in the job specifications. Every effort must be made to produce results that are comparable. To do this, the same equipment, method and when possible the same person should be used at all times throughout the contract. Should the test result be lower than the required strength, then there is something wrong with the concrete or the testing regime. This can be due to any of the following main reasons:

(a) too much water;
(b) insufficient water (this will prevent compaction);
(c) badly made cubes;
(d) badly cured cubes.

The effect of the quantity of water in concrete has been covered in Chapter 8 – Workability.

In relation to cube strength it is sufficient to reaffirm that too much water has eventually to evaporate, leaving behind voids that will weaken the compressive strength of the concrete.

If there is not enough water to enable the concrete to be compacted then air will be entrapped, 1 per cent of air entrapped will reduce the strength of the concrete by about 5 per cent; by leaving 4 per cent of air in the concrete cube its strength can be reduced by around 20 per cent.

Making cubes

Samples of concrete are taken as previously described. The object is to get a fair sample of the total mix and to avoid taking samples from segregated concrete at the beginning of a discharge or from the edges of a discharged heap.

The next step is to ensure that the correct equipment is available and clean.

Equipment

1. Cube moulds and base plates.
2. Spanners.
3. Mould oil.
4. A scraper.
5. A sampling scoop to hold about 5 kg of concrete.
6. A smaller scoop for filling the moulds.
7. A float.
8. A standard tamper bar, 380 mm (15 in.) long, with a ramming face 25 mm (1 in.) square, weighing 1.8 kg (4 lb) or a vibrating hammer or a vibrating table (only on a large contract with a well-equipped laboratory).
9. Cleaning rags.
10. A bucket or barrow for transporting the samples.
11. Damp sacks.
12. Polythene sheeting.
13. Waterproof crayon.
14. A thermostatically controlled curing tank.
15. A waterproof base plate for remixing.

The standard size cube is 150 mm (6 in.) but if the maximum size of the aggregate does not exceed 20 mm (¾ in.) then a 100 mm (4 in.) cube can be used. Only 'purpose-made' moulds from steel or cast iron should be used; skimping or improvising on testing equipment can lead to trouble.

Every care must be taken with assembly, cleaning and storage, otherwise the true strength of cubes will not be obtained.

Cube moulds are usually in two halves that bolt together and have clips to firmly hold down to the separate base plate. They should have had the surfaces oiled after use to prevent rusting. This oil will need to be wiped off and a clean oil film provided on all surfaces, including joints between the two bolted sections and also the base plate. The samples of concrete that may have been taken for slump testing need to be remixed on a waterproof base plate.

The mould, its base plate and bolts must be firm and tight and then filled to about one third full.

Tamp all over, especially in the corners, using at least 35 tamps. The object of this tamping is to remove the air; 35 tamps is the minimum. Very stiff mixes may need more tamping.

The mould is now two thirds full and tamped, then filled to overflowing and tamped again.

(a)

(b)

Fig. 12.6 (a) Using the float to push excess concrete to the centre of the mould
(b) Removing excess concrete so that the mould can be filled flush

Fig. 12.7 Lightly scratching an identifying mark on the finished cube

The surplus concrete is removed and concrete in the mould smoothed over with a float.

While the cubes are wet, identification marks (cube number and date) should be marked with a match or pencil end. The moulds should then be covered with a damp cloth and a polythene sheet and stored inside a room at a normal temperature around 20°C.

Storing cubes

Cubes made the previous day should be taken from the mould, but handled with care and kept warm and wet.

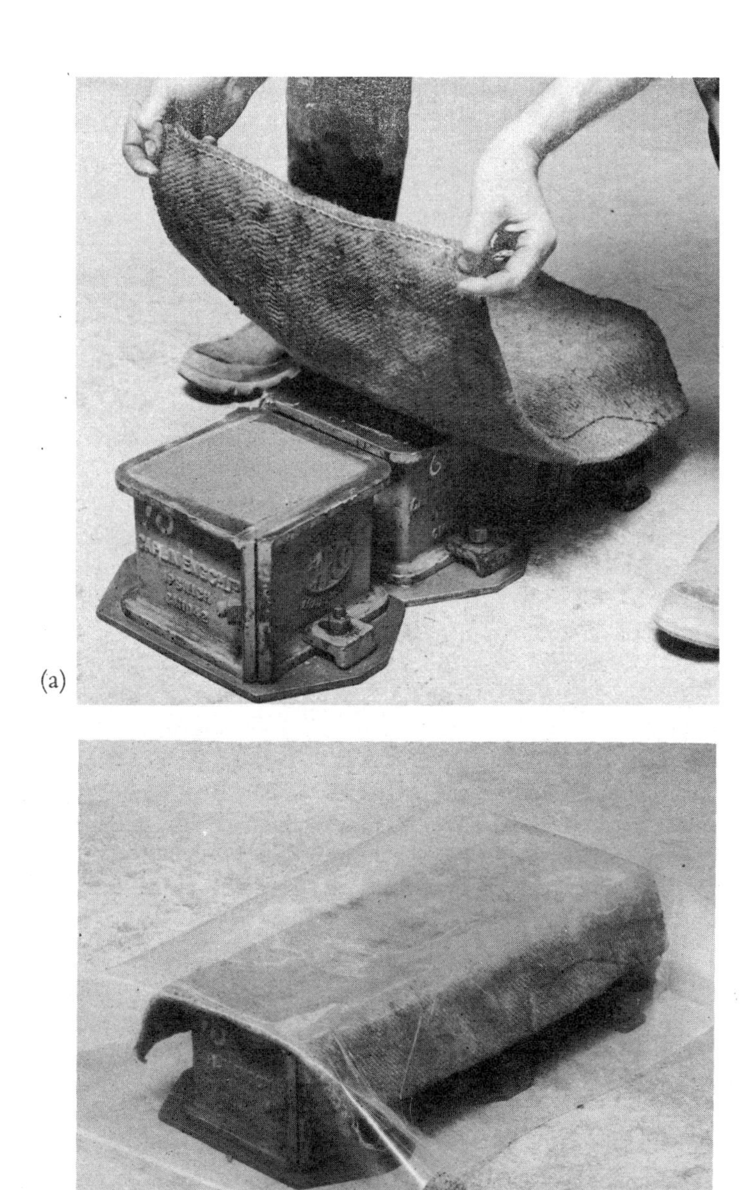

(a)

(b)

Fig. 12.8 (a) Damp sacking for initial curing
(b) Plastics sheeting over the damp sacking

The surplus concrete around the top of the mould is removed with a wire brush. All nuts are then slackened off and the sides of the mould carefully tapped to effect release. The mould should be lifted off gently, taking care not to damage the cube.

Any further marking needed can be done with a waterproof pencil before putting the cubes in the curing tank. The curing tank needs to be large enough to take comfortably all the cubes. The cubes are still soft when placed in the tank and can be easily damaged.

The cubes will only harden when warm and wet. The tank needs to be located to allow regular checking of the thermostat to see that the water remains at 20 °C and that the cubes are covered at all times.

Neglect when curing is a very common cause of low cube results. All the care and effort expended in handling, placing and compacting the concrete can be put in jeopardy if the electricity supply to the curing tank is switched off or the water allowed to evaporate.

Transporting cubes for testing

When the cubes are transported to a testing establishment they will need to be protected against damage and be kept damp. This can be done by wrapping them with wet sacking and placing them in a tied polythene bag. However, if they are to travel long distances then a purpose-made box lined with dampened absorbent material will be required.

The cubes should arrive at the testing laboratory at least 24 hours before the testing time, complete with the appropriate documentation for recording results.

A visit to the selected testing laboratory by a representative of the contractor familiar with the testing procedures is a wise precaution. Recent monitoring of test houses has shown that a considerable percentage are not adhering to British Standard recommendations regarding the regular maintenance of testing machines, training of operatives, storing of cubes, rate of loading and presentation of accurate records.

Any aspect not carried out correctly will result in unreliable test figures, which will reflect adversely on the site's quality control.

It would be regrettable if the effort given to the transporting, placing and compacting the concrete resulted in the quality of workmanship being questioned due to deficiences in testing procedures.

Summary

The quality of concrete is ultimately judged by the appearance and durability of the finished structure.

It is to assess this quality and suitability that rules and methods of procedure have been adopted for control of quality and consistency on site.

The site agent's ability to produce good concrete is largely judged by the recorded results of site tests. It is, therefore, wise to know the rules, to abide by them and to know the reasons behind the manner in which the tests are carried out. To do this one needs:

(a) correct equipment and storage;
(b) staff with ability and knowledge to be able to carry out the tests efficiently;
(c) good records;
(d) to be satisfied that curing, transporting and testing have been carried out to the recommended standard.

Testing is often regarded as a non-productive activity on site, but its cost is negligible compared with the possible cost of having to replace concrete as a result of an incorrectly made test.

Suggested reading

1. Cement & Concrete Association. *Testing for workability*. (Man on the job leaflet).
2. Cement & Concrete Association. *Concrete test cubes*. (Man on the job leaflet).
3. Blackledge, G. F, *The concrete cube test* (2nd edn). Cement & Concrete Association. 1979.
4. Cement & Concrete Association. *Concrete practice*. 1979.

13

Making good and repairs

Good design, careful workmanship and supervision will limit the amount and cost of repairing concrete or making good. Some defects will be inevitable but when these occur remedial work can often be limited to the defective areas. This avoids the costly removal and replacement of otherwise sound concrete.

When the client, or his representatives, permits repairs to concrete, rather than replacement, he is accepting something less than the quality specified. Therefore, the contractor has a duty to supervise properly and to have the repair carried out in the most effective manner. The job cannot be left to an untrained labourer. This will only result in the architect demanding complete replacement on future occasions.

By applying a few basic principles, repairs can be carried out by experienced site personnel, providing advice is sought from an engineer, or a specialist repairer, whenever there is any doubt about the cause of the defect or the safety of the structure.

Causes of defects

Before undertaking any repair it is of utmost importance to find out the cause of the defect, particularly if cracking is present. With structural concrete the advice of an engineer or specialist must be sought before attempting any repair.

Cracking may simply be the result of too rapid a drying out. On the other hand a crack in a concrete wall widening towards the bottom could be due to uneven settlement of the wall foundations or an overload to the top of the wall. In these circumstances it would be pointless to fill the crack before the cause is remedied. In some situations it could be dangerous to do so.

Principles of repair

Cleanliness

It is essential to clean the damaged surface if repairs are to be successful. This can be done in a number of ways, depending on the size and accessibility of the repair.

A multiple scabbling tool can be used for a road surface, a vibrating hammer for a beam, column or wall and even a wire brush and plenty of water on small areas. Grit blasting and high pressure water are other methods.

Once the repair area has been cleaned repair should follow immediately. Delay will allow the surface to become soiled or chemical changes to occur on the surface, thereby reducing bondability.

Materials

Materials identical to those used in the original concrete should be used whenever possible.

Where appearance of the finished concrete is important, trials should be carried out adding white cement or using a lighter colour sand, noting the proportions, carrying out normal curing and leaving the sample for 28 days, preferably in the environment in which it will finally be used.

Techniques

A repair to a vertical face should be finished at right angles to the patch so avoiding 'feathering' out the replacement materials. The replacement concrete in a 'feathered' area will probably be rich in cement and contain too much water. Consequently, it is more likely to shrink and be more susceptible to frost attack and other temperature changes. If the surface is subject to wear then this vertical edge should be at least 6 mm deep.

Operatives must be correctly instructed and have knowledge of how to carry out an effective repair. To achieve this:

1. The right tools must be available; too sharp a chisel for removing defective concrete may split aggregate or damage reinforcement.
2. Sufficient time must be given to complete the repair; patching up by the odd job man between errands is not recommended.

3. Sufficient materials should be available; shortage of the correct materials may lead to substitution, e.g. soft loamy sand instead of sharp sand would seriously reduce the durability of the repair patch.
4. Means and knowledge of correct curing methods must be to hand, together with sufficient resources to carry these out; curing of repairs is essential. The two concretes, the old and the new, will be in different stages of hardening, therefore the moisture must be retained in the new concrete long enough to allow it to gain an equivalent strength.

A record of mix proportions, curing time and any other information should be kept for future reference.

Surface defects

It will be almost impossible to match the repair to the existing concrete in finish, texture and colour; for a variety of reasons the finished repair will generally be darker than the parent concrete.

Where appearance is important the addition of about 20 per cent of white cement to the same proportions of materials that were used in the original concrete will help to reduce this colour difference. However, the making-good material, being applied by hand, will have a different density from that being repaired. Therefore, while replacing defective concrete by patching can be structurally successful, it will always be difficult to disguise its appearance.

Crazing – surface shrinkage cracks – minor blow holes, etc. can be filled by damping the area with a brush and then applying a cement paste with hessian pads – leaving to dry out for a brief period before removing the surplus with dry hessian. Fine silver sand can be added to the grout if filling is considerable. This process is described as 'bagging in' and is often included in the job specification.

Cracking

Care must be taken to distinguish shrinkage cracks from structural cracks. Structural cracks can be filled with specially manufactured penetrating materials subject to the agreement of the structural engineer. These materials are often applied by forming a seal on the face of the crack with adhesive tape and pouring the liquid in a clay or Plasticine cup formed against the concrete at suitable intervals.

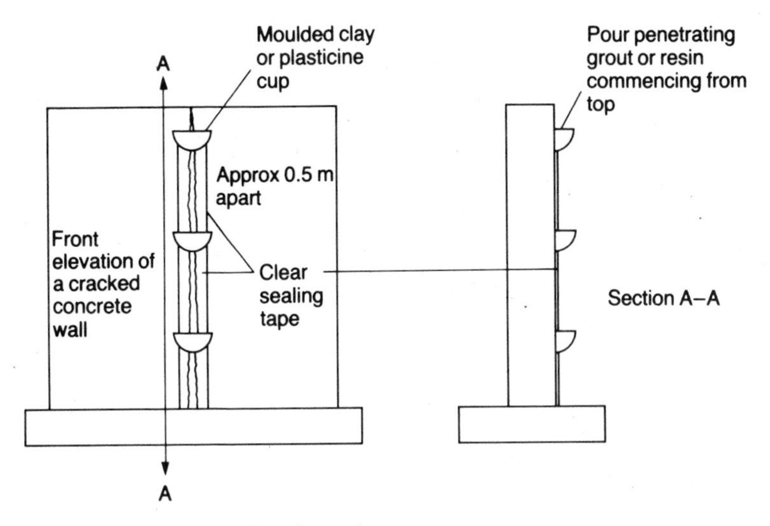

Fig. 13.1 Treatment of wall crack

Replacement of defective concrete with new concrete

We are assuming that an area of wall has been badly compacted and needs repair.

First, agree with all concerned how much needs removing. Unless there are other considerations, the objective will be to remove only the minimum amount of concrete. Poorly compacted concrete should be removed with a blunt tool thereby reducing the risk of damaging the reinforcement or splitting aggregates.

If appearance is important it may be necessary to cut back to the nearest joint or shutter line – if strength is important then one will probably be advised to cut back to enable some additional reinforcement to be attached to the existing steel.

Care must be taken to prevent spalling beyond the chosen area of repair. This can be effected by using a mechanical saw around the perimeter of the patch.

When the cutting away is completed the dust should be thoroughly brushed or blown away and formwork, etc. must be ready to follow on with the repair immediately. If left it will need covering, then thoroughly washing again prior to effecting the repair.

Use the original formwork where possible. New boards or plywood will give a different colour to the finished concrete, necessitating the application of a release agent which should be left to soak in.

Make provision for a hopper or chute to feed in the concrete. Surplus concrete resulting from this will have to be dressed off as soon as formwork can be removed.

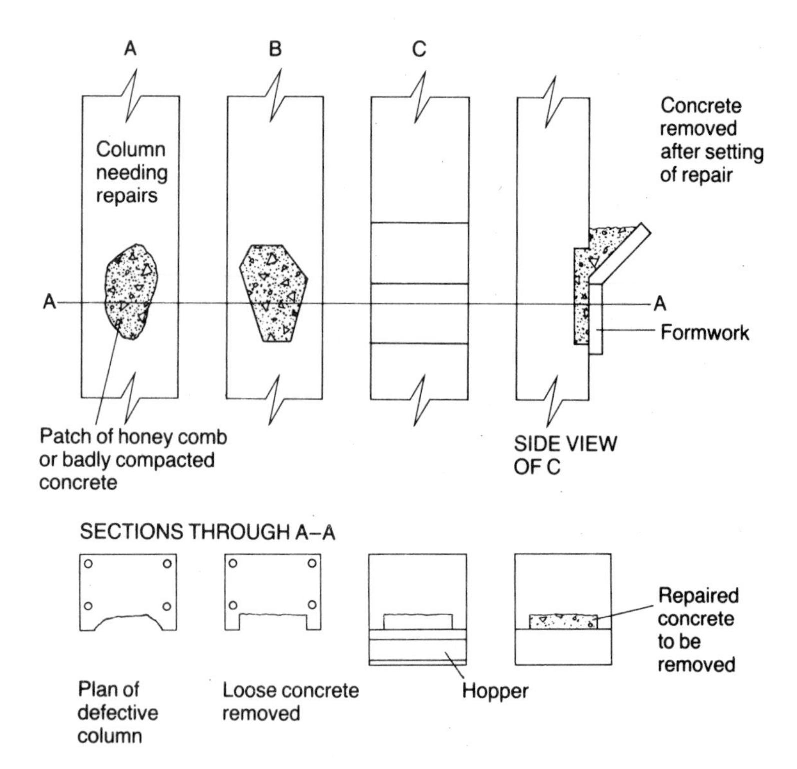

Fig. 13.2 Repairs using 'letter box' or chute

Pressure grouting

Troubles frequently occur at construction joints where poor compaction has taken place around a water bar or at the kicker level.

Excessive cutting away in these circumstances is dangerous, as damage could be caused to the water bar or by disturbing surrounding concrete or hitting the reinforcement with the cutting tool.

A grouting machine can be hired, with or without specialists, which will require strong, well-fixed, shutters to resist the additional pressures. These are fixed and then drilled to accommodate the nozzle of the grout tube; several holes are often drilled and are plugged as fill progresses.

Before erecting the formwork the damaged portion must be treated as previously described. Air pressure and detergents are often used for cleaning to ensure removal of the dust, mould or other oily substances. Grout is then forced in by air pressure to fill all the voids in the honeycombed concrete.

When this method is carried out efficiently it is reasonable to assume that if the repair withstands the pressure of the grout it should be waterproof on completion.

These specialist repairs are very expensive so, if needed, a site meeting should be arranged with the specialists to establish their site requirements in regard to scaffolding, power, materials and assistant labour. As their fees are usually based on travelling time plus an hourly rate on site, all preparation prior to their arrival and assurance of continuity once they commence will help to keep costs to a minimum.

Epoxy resins

Epoxy resins have valuable adhesive properties and are extremely useful for concrete repairs. Their general field of application is:

(a) prevention of moisture penetration through cracked or porous concrete;
(b) repairs to damaged floors or roads;
(c) repairs to precast units, etc.

The usual precautions in preparation must be taken and in addition it is essential that the patch or crack is properly dried and that any aggregate used is also dry.

These resins are marketed in two packs – the resin itself and a hardener. Kept apart they will store for about 12 months but will set rapidly when mixed together. Aggregates can be added to economise on materials, which are usually silica sand 5–8:1 by weight. Pigments can also be added to match colours. Most resin repairs will otherwise be considerably darker than the concrete.

The resins are offered in a variety of forms, depending on the operation to be carried out, so it is wise to discuss fully any repairs with the manufacturer and carefully study instructional leaflets.

Some care is also needed in handling the materials and protective clothing, barrier creams and cleansing agents are necessary.

Injection nipples can be used to repair cracks with epoxy resins temporarily stuck to the face of the crack and a grease gun or similar to provide pressure in application of these resins. Damage can be caused by too much pressure so it is sensible to consult specialists before embarking on this type of repair.

Concrete floor repair with resin materials

There are two main types of materials available for repairing floors; trowelled compounds and self-levelling compounds.

The former are more likely to be required for repairs and are normally used where heavy mechanical wear is expected.

Although these resins can be used to repair floors damaged by acids, etc. they themselves are liable to some attack by organic liquids or inorganic acids, so advice should be sought before commencing this type of repair.

The preparation of floors should be carried out in the normal manner so that they are clean and free of oil or loose materials and they are then treated with a tack coat of resin without aggregates.

Aggregates are then added at a ratio of 5–8:1 (dry and of silica sand, calcined bauxite or granites). Grading is important and further information should be sought from specialists or the Cement and Concrete Association.

The material is placed and then screeded with a steel trowel, the joints in the screed coinciding with those in the structural slab. Curing takes 7 days but light traffic is acceptable after 2 days.

Pigments can be used for matching existing finishes; some hardeners darken with age.

Summary

1. Know what you are doing.
2. Provide all the facilities to carry out the tasks.
3. Seek expert advice when not confident of the remedy.
4. Treat repair operations as seriously as new construction.
5. Having gained permission for the repair it is very important not to lose the architect's trust or goodwill by making a poor job of the replacement work.

6. When you have learnt how to repair it you have gone a long way to prevent it re-occurring in the future.

Suggested reading

1. Shirley, D. E. *Introduction to concrete.* (2nd edn) Cement & Concrete Association. 1980.
2. Cement & Concrete Association. *Making good and finishing.* (Man on the job leaflet).
3. Allen, R. T. L. *The repair of concrete structures.* (3rd edn) Cement & Concrete Association. 1978.
4. Taylor, G. Maintenance and repair of concrete structures. *Maintenance Information Service Paper No.* 16. Chartered Institute of Building. 1981.

14

Surface finishes

Architects and engineers seeking new effects, or economy, require concrete not only to satisfy structural and durability requirements, but also to look decorative, as well as to act as sound barriers and to aid heat conservation.

The structural and durability qualities are well tried and tested, as has been demonstrated in earlier chapters. These qualities can be measured and controlled, but decorative finishes can prove more difficult to produce. Due to the plastic nature of concrete it can be moulded into patterns and shapes that are easily reproduced in a small scale. However, defects often occur when dealing with large areas due to the complexity of the interrelationships of all those concerned with the concrete, i.e. from architect to site agent.

The site agent must understand the requirements of the designer and, what is more important, he must be sure that they can be met and that he can meet them.

Concrete finishes are often poorly specified and in extreme cases impossible to achieve. The site agent should be wary of a specification that demands a 'fair face' finish or uses loose indefinable phases such as 'to the architect's satisfaction'.

The current Code of Practice has recommendations defining a variety of surface finishes that are helpful to both designer and site agent.

Where possible the contractor should ask to see an example of the finish, preferably a building, that he is asked to price and to produce. A surface finish based on a small pre-cast sample may cause considerable problems when scaled up. Placing concrete in large quantities, perhaps taking several hours, is likely to be marred by 'layering' due to successive batches of concrete not being uniform in colour. The solution to this particular problem may be to change the method of construction by pre-casting the concrete in smaller panels under controlled conditions.

Colour

To produce facing concrete to a consistent colour on a large contract demands very rigid control indeed.

A satisfactory result is dependent on a number of factors and some are beyond the site agent's control.

When uniform colour is important, aggregates, particularly the sand, must be supplied from only one source and even then, if the contract is continuing over a longer period, the materials should be extracted from one part of the source and stockpiled in sufficient quantities to meet the total requirements.

Very strict control will also be essential to ensure that the mix proportions and mixing times are consistent. The moisture changes will need to be carefully monitored and adjusted as needed.

Storage of aggregates away from any sources of contamination is also important; surplus tea or coffee drainings can alter the colour.

Formwork

Formwork is undoubtedly next in importance. The concrete will mirror every surface detail and quality of finish will, therefore, depend largely on the design and quality of the forms.

Providing the selected formwork surface is obtainable in sufficient quantities to maintain consistency, the re-use factor will determine the choice of material (e.g. with care plastic faced plywood will provide more re-uses than softboard finishes).

Formwork design is critical; the forms must be watertight as any leakage will alter the ratio of water to cement and the distribution of the fine particles. This results in dense dark blemishes.

Compressible foamed polyurethene is useful to seal formwork joints.

The difficulty of fixing foam strips to plastic-coated plywood can be solved by forming a saw-cut at the junctions and pressing in the foam (see Fig. 14.1).

This same attention to joint detail must also be employed. even when the surface is to be removed to leave an exposed aggregate finish. Otherwise these blemishes will still be visible due to the collection of 'fines' at the leak.

The forms must be able to resist the pressures without distortion and to accomplish this the rate of pour and method of placing

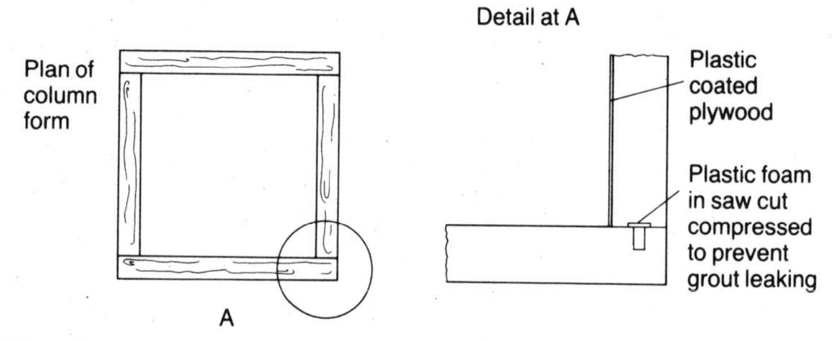

Fig. 14.1 Method of grout leak prevention when using plastic coated formwork

may have to be modified. However carefully the formwork is designed and the placing of concrete carried out, the junctions between hardened and fresh concrete will be visible and unsightly unless special provision is provided to disguise these joints. This is usually accomplished by providing a half round or dovetailed feature, half of which is cast in the first concrete and then refitted and completed in the following lift. (see Fig. 9.2).

Consistent and sufficient vibration is equally important, particularly when using dense form materials. For example, with plastic-faced plywood, the likelihood of unreleased air bubbles causing voids on the surface is greater than with sawn boards. Increased propping and strutting will probably be needed to reduce deflections, acceptable in ordinary structural concrete, but which will mar the appearance of decorative surfaces.

Finally, the care taken in cleaning after striking and in storing will make a significant contribution to success or failure.

The application to the forms, by spray or brush, of a light coat of the release agent used on the project will provide the protection needed.

Careful stacking to prevent distortion and clear marking to enable re-use in the sequence adopted for re-erection will extend the useful life of the forms and effect economies in handling.

Vibration

Concretes having a low water/cement ratio need mechanical vibration to aid compaction. However, this vibration can also contribute to unwanted surface blemishes.

The vibrations of the form face sometimes produce variations in the density of the concrete, particularly when using very smooth plastic-faced plywood. This shows as dark patches or shadings on the finished surface. These are permanent and cannot be easily disguised.

Vibration to a mix that is likely to bleed will bring the water to the surface and this will show scouring and also different colouring to the finished concrete.

If the formwork is not sufficiently robust the vibration will reveal the joint and construction weaknesses, causing leaking of cement paste that leave blemishes on the surface.

Curing

The purpose of curing decorative surface finishes on columns, beams and walls is to prevent rapid loss of moisture. Should forms be struck too soon the exposed concrete will be lighter than concrete which has remained covered and has retained the moisture. To assist moisture retention the tops of columns and walls should be covered after at least 4 days or longer in cold, winter conditions. Similar curing and working conditions for each day's work should be sought.

Secondary curing – wet hessian or applied curing agents are not really necessary on vertical surfaces in the UK and can often cause staining or discolouration. It is difficult to apply these methods uniformly.

Profiled or textured finishes can only be produced satisfactorily by strict attention to uniformity of materials and supervision of use and storage of formwork. An adequate stock of forms should always be available.

Variations in colour that occur when repairing a form can be minimised by keeping a reserve of replacement formwork to be used when required.

This formwork will need oiling for several successive days before use to seal the pores. Only then will it produce a surface comparable to that previously formed.

Applied finishes

Concrete that has to receive any applied finish must be clean and sound. The surface of formed concrete is obtained by bringing the water and fine particles of cement, sand, air and any fine impurities to the surface. Left for a period this surface also collects dust and impurities from the atmosphere.

Washing the surface, however well done, is not the best method of achieving a bond. Therefore, for the successful application of a finish, this surface must be removed, by mechanical means or by use of retarding agents.

Once removed the area should be washed and the finish applied as soon as possible, preferably while exposed aggregate is damp. Clean surfaces must not remain exposed to collect site dust.

Improved communication

The production of a satisfactory building in decorative concrete requires the blending of the designer's skills with the sound planning and site experience and site organisation from the contractor, together with the skills and knowledge of his site staff.

Improved communication and full co-operation are necessary to enable achievement of consistent high quality.

Summary

Endeavour to visit a building incorporating an example of the finish that has been specified.

Large areas of uniform colour are difficult to produce. They require good planning of material delivery and close supervision of workmanship. Formwork, quality and consistency and care become more critical when producing decorative finishes. Stocks must be ample to provide for replacements.

Design must prevent leakage and deflection. Consistency is also the keynote for vibration; variations in application will be visible in the appearance of the finished surface.

Indifferent attention to curing times and method will be revealed by having varied surface finishes; forms will probably have to be left in position longer after pouring to obtain similar finishes.

Cleanliness is essential when applying a finish to a concrete. This is usually achieved by removing the formed surface to expose the aggregate. Co-ordination between all parties is vital to the success of a decorative concrete structure.

Suggested reading

1. Monks, W. *Visual concrete – design and production*. Cement & Concrete Association. 1980.

2. Monks, W. & Ward, F. *External rendering*. Cement & Concrete Association. 1980.
3. Wilson, J. G. *Specification clauses covering the production of high quality finishes in in-situ concrete*. Cement & Concrete Association. 1970.
4. Cement & Concrete Association. *Tooling concrete*. (Man on the job leaflet).
5. Baker, E. M. Workmanship factors associated with CP110 – The structural use of concrete. *Site Management Information Service Paper No*. 88. CIOB. 1981.
6. Cement & Concrete Association. *The control of blemishes*.

Index